W0091441

Industrial Polymer Applications

Essential Chemistry and Technology

Industrial Polymer Applications

Essential Chemistry and Technology

William R. Ashcroft

THE QUEEN'S AWARDS
FOR ENTERPRISE:
INTERNATIONAL TRADE
2013

Print ISBN: 978-1-78262-814-9

A catalogue record for this book is available from the British Library

Published by The Royal Society of Chemistry,
Thomas Graham House, Science Park, Milton Road,
Cambridge CB4 0WF, UK

Registered Charity Number 207890

Visit our website at www.rsc.org/books

Printed in the United Kingdom by CPI Group (UK) Ltd, Croydon, CR0 4YY, UK

Preface

Almost every type of thermoplastic and thermoset polymer has found use, at some time following their discovery and development, in the modification, protection, repair, restoration and bonding of industrial engineering materials. This book considers why resins, polymers and plastics are needed, how they are used, routinely or occasionally, and describes the specific problems and relevant industrial application challenges that can be overcome through their use.

The various chapters offer a systematic and balanced view on the wide diversity of properties and applications for well-known and less well-known thermoset resins, prepolymers, polymers and thermoplastics made use of with each of the main classes of engineering materials in turn, starting with concrete, masonry, wood, metal, rubber, plastic, and ending with glass and advanced ceramics. Where and how rudimentary commercial, as well as more sophisticated thermosets and thermoplastics compare, along with fundamental fitness-for-purpose testing which defines and validates them, is described application by application. Supporting information in the form of "Fact File" boxes are included throughout to guide the reader through the terminology and jargon, delineation, purpose, peculiarities and subtleties associated with engineering materials and polymers as well as their verification and endorsement where pertinent. "Application Challenge" boxes are an integral feature of the text dealing with commonly encountered natural or inherent phenomena that may be surmounted by judicious industrial polymer applications.

The objective of Industrial Polymer Applications is to provide a supplemental text for undergraduates, postgraduates and industrialists who are studying or are involved in the physical sciences and

Industrial Polymer Applications: Essential Chemistry and Technology
By William R. Ashcroft
© William R. Ashcroft 2017
Published by the Royal Society of Chemistry, www.rsc.org

engineering. This new book brings together and presents a balanced appraisal of the capabilities for the most notable and likely polymer chemical technology options with industrial applications to complement existing books, published articles and patents written specifically about individual polymer technologies. As such, the text and accompanying appendices are not intended to answer every question there may be about the chemistry, technology, testing or industrial application challenges which can be overcome by the use of resins, prepolymers, polymers and thermoplastics, nor does it bid to tabulate physical/mechanical property data for every individual commercial option—as a deliberate alternative, diagrams are used to accentuate comparative characteristics and performance of polymer options.

Recommended reading suggestions are routinely included where more detailed knowledge or understanding of polymer characteristics, engineering material usage and testing may be required. These are intended to provide focus where specific product/solution recommendations are required and internet searches may reveal an overwhelming choice, or offer conflicting recommendations and unclear information about actual suitability for an intended application. The suggested reading is also intended to present balance in cases where trade association webpages, blogs and publications with business sponsors, or private company webpages which are naturally inclined towards particular commercial and technological strengths, may without knowing miss out all viable options.

Only essential chemistries and technologies are included in this book for ease of assimilation and comprehension of the capabilities and suitability of industrial polymers and their applications. Reference information about the generic composition, structure, characteristics and uses for each of the resin, polymer or plastic types referred to in the text is provided as a glossary, along with a separate appendix containing worked examples for stoichiometry calculations for thermoset-curing resin technologies. There is also an appendix listing all the fundamental international standards, practices, specifications and test methods cited throughout the book.

William R. Ashcroft

Acknowledgements

I would like to express my gratitude to all my instructors, employers, former colleagues and business associates who have educated, nurtured, helped and guided me during a long and gratifying career. I am also indebted to the editorial staff of the Royal Society of Chemistry for accepting and transforming the manuscript.

Industrial Polymer Applications: Essential Chemistry and Technology
By William R. Ashcroft
© William R. Ashcroft 2017
Published by the Royal Society of Chemistry, www.rsc.org

Dedication

To Alison and Helen for their endless encouragement, support and help throughout the preparation of this book, and for their patience and understanding in previous years.

Industrial Polymer Applications: Essential Chemistry and Technology
By William R. Ashcroft
© William R. Ashcroft 2017
Published by the Royal Society of Chemistry, www.rsc.org

Contents

Industrial Polymer Applications: Essential Chemistry and Technology
By William R. Ashcroft
© William R. Ashcroft 2017
Published by the Royal Society of Chemistry, www.rsc.org

1 Concrete Modification, Protection and Repair

1.1 Introduction to Concrete and the Need for Modification, Repair and Protection

Concrete is the earliest known synthetic engineering composite and is characterised by its ease of use and ability to be formed into a multiplicity of shapes which harden over time into a durable rock-like material. It is the most versatile and widely handled construction material in building and civil engineering projects, since it can be used with or without reinforcement, because it can be pre-stressed or post-tensioned, and as it can be chemically modified in a variety of ways to boost physical and mechanical properties as well as to meet specific application requirements.

Unreinforced concrete is the best-known example of a large-particle reinforced composite and is made from blends of gravel and sand aggregate embedded in a matrix of Portland cement paste – sand is required to pack into the gravel voids so the aggregate blend provides solid reinforcement to the cement matrix to improve mechanical strength; cement acts as the binder which holds the overall mixture together. Combinations of cement, water and only fine sand aggregates, referred to as mortars, perform in exactly the same way as bulk concrete whereby the principal components of the cement react with water (eqn (1.1) and (1.2)) to yield a calcium silicate hydrate (tobermorite) solid gel. The solid tobermorite gel grows in the form of fibrils, which ultimately interlock to form a

Industrial Polymer Applications: Essential Chemistry and Technology
By William R. Ashcroft
© William R. Ashcroft 2017
Published by the Royal Society of Chemistry, www.rsc.org

continuous solid matrix coating around and bonded to the aggregate particles.

$$2CaO \cdot SiO_2 \text{ (dicalcium silicate)} + xH_2O \gg Ca_2SiO_4 \cdot xH_2O \text{ (tobermorite)}$$

$$(1.1)$$

$$3CaO \cdot SiO_2 \text{ (tricalcium silicate)} + (x+1)H_2O \gg Ca_2SiO_4 \cdot xH_2O + Ca(OH)_2$$

$$(1.2)$$

There are differences between the two calcium silicate minerals on reaction with water – the dicalcium silicate ($2CaO \cdot SiO_2$) sets more slowly but produces higher strengths, and the tricalcium silicate ($3CaO \cdot SiO_2$) sets rapidly but produces low strengths. Both generate heat, which can cause loss by evaporation of water which is required to ensure complete hydration. The unavoidable presence of $3CaO \cdot Al_2O_3$ (tricalcium aluminate) created during clinkering gives rise to the phenomenon of "flash set", generating a large amount of heat of hydration unless additions of $CaSO_4 \cdot 2H_2O$ (gypsum) and CaO (lime) are made to divert reaction to form an insoluble layer of $3CaO \cdot Al_2O_3 \cdot 3CaSO_4 \cdot 32H_2O$ (ettringite) over the surface of the aluminate crystals and subsequent slow reaction to form the $3CaO \cdot Al_2O_3 \cdot CaSO_4 \cdot 12H_2O$ hydrate which reduces the overall durability of concrete.

The rate of set and final strength of concrete and cement-based mortars is not only controlled by the water content and the relative amounts of the various slow and fast hydrating minerals in the cement, but also by addition of accelerators which increase the rate of hydration and retarding agents which are used to slow down/control the set time.

> ### Fact File: Portland Cement
>
> Portland cement is produced from finely inter-ground mixtures of the various minerals of calcium which result from clinkering limestone and clay through a rotary kiln in the historical Portland process. Typically this kiln product, called cement clinker, is composed primarily of $2CaO \cdot SiO_2$ (dicalcium silicate) and $3CaO \cdot SiO_2$ (tricalcium silicate), with lower amounts of $4CaO \cdot Al_2O_3 \cdot Fe_2O_3$ (tetracalcium aluminoferrite), $3CaO \cdot Al_2O_3$ (tricalcium aluminate) and $CaSO_4 \cdot 2H_2O$ (gypsum). The kiln product is then ground with varying amounts of gypsum to produce the various types of commercial Portland cements covered in the BS EN 197 and ASTM C150 specifications:
>
> Type I – general use cements when the special properties specified for the other types are not required;

Type IA – air-entraining cement for the same uses as Type I, where air-entrainment is desired;

Type II – for general use, more especially when moderate sulfate resistance is desired;

Type IIA – air-entraining cement for the same uses as Type II, where air-entrainment is desired;

Type II(MH) – for general use, more especially when moderate heat of hydration and moderate sulfate resistance are desired;

Type II(MH)A – air-entraining cement for the same uses as Type II(MH), where air-entrainment is desired;

Type III – for use when high early strength is desired;

Type IIIA – air-entraining cement for the same use as Type III, where air-entrainment is desired;

Type IV – for use when a low heat of hydration is desired;

Type V – for use when high sulfate resistance is desired.

The differences between the various types are rather subtle and have developed over time to meet specific performance requirements in different concrete and cement-based construction applications. In addition, white cements which contain no more than 0.50 wt% of ferric oxide (Fe_2O_3) are available and are used to ensure clean bright consistent colours for aesthetic architectural concrete as well as masonry and cementitious products.

Why the need for polymeric modification, repair and protection? There are a variety of issues inherent to concrete manufacture and deterioration in service which call for repair, maintenance and even overhaul solutions to conserve functional integrity. Problems start with the water: cement ratios and water levels used in manufacture, which have a marked influence on the workability of concrete and mortar mixes – to obtain a workable combination, more water is required than that which is needed for hydration. Unfortunately, any excess water lowers the density and creates voids in the set concrete, which contribute to shrinkage cracking and reduction of overall durability and create an ongoing need for maintenance, repair and overhaul starting from the point of casting. Inclusion of water-reducing agents, air-entraining agents, and polymer modifiers (see Section 1.2) in freshly mixed concrete can, however, permit equal workability at below-normal water levels with relatively little reduction in final strengths.

Fact File: Concrete Reinforcement

When reinforcing bars (rebars) and mesh made from high tensile and compressive strength steel, polymers or alternate composite materials are embedded into concrete before it sets, the resulting reinforced concrete is able to withstand applied stresses (tensile, bending and compression). Incorporation of fibrous materials made of steel, glass, synthetic and even natural fibre also increases structural integrity, depending on their length, distribution and orientation with, for example, lightweight thermoplastic fibres in 1- and 2-component solid premixes providing a simple alternative to embedding light reinforcement mesh.

It is evident that concrete is an inherently complex material to produce and, somewhat frustratingly, it also proves difficult to repair as new concrete, cementitious mortars and grouts will not stick well to old/new concrete without polymer modification or without the use of polymeric bonding, priming or conditioning agents. Concrete is not only prone to shrinkage cracking as it matures, it has essentially low tensile strength and elasticity, which decreases under tension causing cracking within the composite matrix, and even though it has relatively high compressive strength long-term structural loading can cause concrete to deform or creep.

The inherent porosity and alkaline nature of concrete also make it susceptible to liquid ingress as well as biological and chemical attack (even from carbon dioxide from the air – see box) if left unprotected. The following sections provide an insight into the essential chemistries and technologies of the various polymer types used to supplement or substitute cement as a binder in the design and use of concrete replacement, as well as for repair and protection.

Application Challenge – Concrete Carbonation

Carbon dioxide from the air can react with the calcium from calcium hydroxide and calcium silicate hydrate in concrete to form calcite ($CaCO_3$) in a process known as carbonation. The surface of fresh concrete reacts quickly, and the more porous and permeable the concrete, the faster and deeper the penetration inside into the concrete pore fluid. Although the process of carbonation increases both the compressive and tensile strength of the concrete, there is a resulting decrease in alkalinity which leads to problems of corrosion for embedded steel reinforcement – when the pH falls from the normal 12.5–13.5 range encountered in concrete pore fluid before carbon dioxide penetration, and whilst it remains at pH 10 or above, steel reinforcement is passivated and protected from corrosion; when, however, the pH drops below 10, the steel's thin layer of surface passivation dissolves and corrosion occurs. Pre-coating of steel reinforcement or replacement with non-corroding reinforcements are two options to prevent the corrosion process; sealing of the pores or application of a protective coating to the surface are other options (Sections 1.4 and 2.2). When left unaddressed, the corrosion of the reinforcing steel can lead to spalling of the embedding concrete, which in turn leads to accelerated deterioration of structural integrity – it will then be necessary to make repairs with polymer concrete mortars (Section 1.3).

Recommended Reading

- M. Neville, *Properties of Concrete*, John Wiley & Sons Inc., 5th edn, 2012, ISBN-13: 9780273755807.
- N. B. Winter, *Understanding Cement*, WHD Microanalysis Consultants Ltd., 2012, ISBN-13: 9780957104525.
- M. Raupach and T. Büttner, *Concrete Repair to EN 1504: Diagnosis, Design, Principles and Practice*, CRC Press, 2014, ISBN: 9781466557468.
- *Concrete Repair: A Practical Guide*, ed. M. G. Grantham, CRC Press, UK, 2014, ISBN: 9780415447348.
- M. H. Irfan, *Chemistry and Technology of Thermosetting Polymers in Construction Applications*, Springer Link, 1998, ISBN: 9789401060790.

1.2 Polymer-modified Cement Repair and Restoration

This section describes the major industrial repair and restoration applications for polymer-modified cement concretes, mortars and levelling/wearing screeds – other applications such as adhesives, grouts and waterproofing are covered in subsequent sections.

Fact File: Polymer-modified Levelling and Wearing Screeds

BS 8204, ISO 6707 and EN 13813 refer to a levelling screed as "a screed suitably finished to obtain a described level and to receive the final flooring". A wearing screed is defined as "a screed that serves as a flooring". A pumpable self-smoothing screed is defined as "screed that is mixed to a fluid consistency, that can be transported by pump to the area where it is to be laid and which will flow sufficiently to give the required accuracy of level and surface regularity". This is often referred to as a "self-levelling screed" and is also used as a levelling screed or a wearing screed.

1.2.1   Essential Chemistry and Technology

Portland cement mortars and concretes are typically modified with thermoplastic polymer emulsions or powders to improve the workability of mixtures at reduced water cement ratios, thereby increasing the density of the ultimate finished particulate composite and reducing the tendency for shrinkage and cracking. Polymer emulsions, known as gauging liquids, are supplied as part of a two-part system to keep then separate from the solid hydraulic cement component; powdered polymers are supplied pre-mixed in the solid hydraulic cement component as one-part alternatives to which water is added at point of use.

As water is lost during the curing process of a cementitious polymer concrete mix, suspended polymer particles coalesce to form an interpenetrating organic polymer network which interlaces with the inorganic network of hydrating cement and embedded aggregate. During the early stages of cure, coalesced thermoplastic polymer deposits function as a physical barrier which helps trap and retain levels of internal moisture sufficient to ensure complete hydration of the cement. In comparison with unmodified concrete particulate composites, polymer-modified cement composites are characterised not only by their ultimate enhanced toughness and resilience, but also by their improved adhesion to properly prepared and conditioned old concrete or masonry substrates, as well as for overcoating with bonded polymeric membranes, coatings or resin flooring.

There are numerous generic types and grades of solid and emulsion polymer modifiers used to modify cement, of which acrylics provide the best resistance to chemical breakdown on exposure to sunlight and temperature extremes. Styrene-butadiene rubber co-polymers (SBRs) are used for maximum acid/alkali resistance, but they are susceptible to discolouration and degradation on exposure to ultraviolet light. Styrene-acrylic co-polymers (SAs) are used where enhanced strength in thin sections is required. Polyvinyl acetates (PVAs) are the most cost-effective additives but vinyl acetate ethylene (VAE) copolymers offer a number of performance advantages over PVAs including increased thermal stability, improved tack and adhesion at low temperatures as well as better flexibility and water resistance which is due to reduced plasticiser requirement. Waterborne epoxy dispersions are used where adhesion and an effective barrier against carbonation and penetration of water and chlorides is required. Figure 1.1 presents a qualitative

Figure 1.1 Environmental degradation resistance comparison of emulsion polymer modifiers.

assessment[†] of the resistance to degradation – on exposure to sunlight, water and high temperatures – of these key emulsion polymer modifiers.

Particulate aggregate options for polymer-modified concretes involve two principal size groups: coarse aggregates of more than 5 mm (200 thou, mil) in diameter such as crushed stone and chippings, as well as gravel; and fine sand aggregates which are less than 2.5 mm (100 thou, mil) in diameter. The different particle size groups are required and deliberately blended together to maximise packing and minimise the void volume – this is discussed in more detail in Section 1.3. In all cases, angular rather than round aggregates are preferred for their ability to mechanically interlock and for their greater surface area for bonding, attributes which both contribute to the development of higher strengths in cured concrete. Extremely fine aggregates, although useful in principle for void filling and minimising porosity in finished concretes, have a detrimental effect on water, polymer and cement demands owing to adversely high surface areas, so their presence generally has to be limited to ensure finished concretes are not over-extended/left with insufficient cement to bind all of the aggregate.

Aggregate selection is customarily influenced by the requirements of the application as both the nature and particle size distribution have a significant impact on aesthetics, costs, strength and density in particular, which changes depending on the amount, density and maximum size of the aggregate used, as well as the cement concentration and how much air is entrained. Polymer-modified and unmodified normal strength Portland cement concrete densities range from 2240 to 2400 kg m^{-3} (140–150 lb ft^{-3}), whereas high-density concretes range from 3040 to 4160 kg m^{-3} (190–260 lb ft^{-3}) and lightweight concretes range from 1440 to 1840 kg m^{-3} (90–115 lb ft^{-3}). Heavier concretes are made by incorporating dense materials such as metal shot for radiation containment (absorption) applications; lighter concretes incorporate mineral slag by-products from the

---

[†]This simplified diagramtic approach to the comparison of the properties or characteristics of different types of materials is used throughout this text as an alternative to representations typically adopted by others *e.g.* P. A. Dunn, *Polymers Paint Colour Journal*, 1984, Aug. 8/22.

steel-making and steel-refining processes, expanded glass or even polystyrene beads for improved thermal insulation applications in addition to weight reduction.

Aggregate purity and moisture contents are also important so mineral aggregates are generally pre-washed and dried to reduce the presence of excess water and, more importantly, reduce any contamination with inorganic salts that can contribute to the development of efflorescence in concrete – a phenomenon which also affects masonry and is discussed in detail later, in Chapter 2.

There are also numerous other types of additives used routinely in the formulation of polymer-modified cement concretes, mortars and levelling/wearing screeds. They are included to facilitate workability and air entrainment, increase or retard the rate of hydration of cement, enhance the tack/grab, or thicken in order to improve slump resistance or permit inclusion of heavier aggregates.

1.2.2   Fit-for-purpose Reactivity Testing

A good working life of a bucket mix of polymer-modified concrete, mortar or screed is important to avoid wastage, but more important is the time the mix applied in a repair area takes to set and harden and allow access for foot traffic. There is no especially good cure time test for thin sections or scrim coats of cementitious renders, but ASTM C191/BS 4550 is a simple and reliable method for the determination of the setting times of thick section repairs. It requires freshly mixed samples to be placed in a set volume rigid conical mould and the penetration of a weighted needle measured every 10–15 minutes. The initial set time is determined from the time to reach a penetration value of 25 mm or less; the hard dry time is the time after which the needle no longer penetrates the sample.

Prior to subsequent application of an impermeable functional protective overcoating (see Sections 1.4 and 1.5) the moisture content of any hydrated cementitious repair material needs to be sufficiently low to avoid any interference with any following surface treatment. When there are doubts about the moisture level in the concrete repair, the simple *in situ* test method ASTM D4263 can be used. This involves taping a plastic sheet to the surface of the concrete repair and using a dew point hygrometer to test the level of moisture in the air under the sheet to confirm how much evaporation is occurring over the course of 72 hour time periods. It is clearly safe to overcoat once moisture has ceased to be released.

Application Challenge – Green/Damp Concrete

Excess moisture, present in green or damp concrete over which an impermeable surface treatment is placed, will become trapped under any impermeable sealer or top-coating. Over time, hydrostatic pressure will force trapped moisture upwards and cause bubbles in a surface treatment or cracks in any impermeable coating applied over it. Unfortunately, the process of evaporation of excess moisture from a repair into the air – so the cementitious concrete can dry and harden – is a process that can take days or even weeks. This is due to the speed of evaporation being determined by the ambient temperature and humidity of the surrounding air. As long as the vapour pressure in the concrete is greater than that in the air, water will continue to evaporate from it. However, if excess moisture is present in the repair when an impermeable surface treatment is installed, it will become trapped under that covering or seal. The ideal time to install an overcoating is when the vapour pressure between the two surfaces is in equilibrium, and unfortunately no laboratory procedure exists which can predict a suitable overcoating window for variable temperature and air humidity conditions. Fortunately, there is a straightforward practical application solution to minimising potential problems, involving good ventilation/airflow with or without space heating over the repair to facilitate drying. Alternatively, epoxy-modified levelling screeds can be used as a surface damp-proof membrane to allow for the early installation of impermeable coatings and floor finishes.

1.2.3   Fit-for-purpose Performance Testing

Trowel-applied and flow-applied screed floorings designed as wearing surfaces require a number of specific adhesion, mechanical strength and physical property testing regimes.

Fit-for-purpose adhesion testing by direct tension (pull-off) methods for polymer-modified mortars on concrete results generally in substrate failure as bond strengths ($\geq$1.5 MPa) are characteristically greater than the cohesive strength of C25/30 concrete. In practice, adhesion in tension, flexure and compressive shear are proportional to the polymer:cement ratio, regardless of the type of polymer tested, with adhesion reaching an optimum at a polymer:cement ratio of about 5% by weight.

Fact File: Concrete Grades

Concrete is typically characterised by compressive strength, measured 28 days after pouring and C25/30 concrete refers to unreinforced concrete with a 150 mm diameter$\times$300 mm cylinder compressive strength value of

25 MPa and a 150 mm cube compressive strength value of 30 MPa as determined by BS EN 12390. This is the grade typically encountered in most industrial applications with heavy-duty construction projects using C40/50 rated concrete (40 MPa cylinder compressive strength/50 MPa cube compressive strength). In the USA, commercial work requires concrete with compressive strength of 4000 psi (27.5 MPa) or greater and 6000 psi (41.4 MPa) or higher for heavy-duty commercial applications.

Cementitious levelling and wearing screeds are typically characterised on the dual basis of compressive and flexural strength (Table 1.1)[‡] with requirements of ≥C20 and ≥F5 strength class for standard, medium duty screeds of thickness 10–25 mm.

There is a variety of methods for the determination of compressive strength including EN 13892 and ASTM C873 which verify fitness-for-purpose by crushing cylindrical "pop-out" cast specimens in a Universal Testing Machine in a direct measure of the yield compressive strength – unmodified concretes typically exhibit compressive strength values in the range of 20–70 MPa (3000–10 000 psi) depending on exactly how they are formulated, whereas polymer-modified concretes generally give lower values in the range 10–60 MPa (1500–8500 psi), which are also dependent on how they are formulated for workability. Testing of specimens that are cast in place using moulds not only provides a measure of the load-bearing capacity of the repair materials, but also the time needed for formwork removal, and the effectiveness of curing where the depth of concrete is from 125 to 300 mm for repair applications.

Flexural strength measurement, according to EN13892 and ASTM C78 methods involves loading a cast beam specimen having either a circular or rectangular cross-section in a Universal Testing Machine in the ISO 178 three-point flexural test technique to determine the highest stress experienced within the material at its moment of rupture. The European Federation of National Associations Representing

Table 1.1   Strength classes of cementitious levelling and wearing screeds.

	Cementitious levelling and wearing screeds characterisation											
Compressive Strength, MPa	5	7	12	16	20	25	30	35	40	50	60	70	80
Flexural Strength, MPa		1	2	3	4	5	6	7	10	15	20	30	40	50

[‡]Mortar Industry Association (part of Mineral Products Association) Data Sheet 22, Issue 1, May 2012 identified from *BS EN 1381, Screed material and floor screeds. Screed material. Properties and requirements*, BSI, 2002, ISBN: 0580407101

producers and applicators of specialist building products for Concrete (EFNARC) also identify elastic modulus in flexure, determined from the slope of the stress-strain curve produced in a flexural strength test, to check the tendency for a material to bend under load.

The physical properties that confirm the likely performance in service of polymer-modified cementitious flooring are wear resistance and skid resistance. EN 13892 prescribes a number of methods for determining the wear resistance (also known as abrasion resistance) of *in situ* floor surfaces, and ASTM C779 also prescribes methods using different machines, each of which apply a different abrasion mechanism to simulate different types of abrasive action. ASTM D4060 represents the more commonly cited laboratory method for testing screed materials wherein cast/cured discs are rotated against a pair of abrasive wheels under a set force such that a circular track is worn away. The test can be performed under both wet and dry conditions, but to emulate heavy traffic/underfoot wear it is the dry sliding method using softer CS-17 wheels that is preferred for validation of fit-for-purpose. Skid resistance of roadways and slip resistance of pedestrian walkways is measured in the EN 13036-4, BS 7976-2 and ASTM E303 pendulum test methods, as well as by numerous other standardised methods involving proprietary equipment which determine dynamic and static coefficients of friction (CoF) values for wet/dry surfaces (see Section 1.5 for details).

Recommended Reading

- R. Bedi, R. Chandra and S. P. Singh, *J. Compos.*, 2013, **2013**, 948745. doi:10.1155/2013/948745 [Mechanical Properties of Polymer Concrete].
- Y. Ohama, *Handbook of Polymer-Modified Concrete and Mortars*, William Andrew Publishing, 1995, ISBN: 978-0-8155-1358-2.
- M. Miller, *Polymers in Cementitious Materials*, Smithers Rapra Press, 2008, ISBN-13: 978-1859574911; ISBN-10: 1859574912.
- Y. Ohama, K. Demura, H. Nagao and T. Ogi, *Adhesion Between Polymers and Concrete*, Springer Link, 1986, pp. 719–729, ISBN: 978-0-412-29050-3 99.
- *ASTM C33, Standard Specification for Concrete Aggregates*, ASTM International, West Conshohocken, PA, 2013.
- *ASTM C1438 – 13, Standard Specification for Latex and Powder Polymer Modifiers for use in Hydraulic Cement Concrete and Mortar*, ASTM International, West Conshohocken, PA, 2013.

- www.efnarc.org the homepage of the European Federation of National Associations Representing producers and applicators of specialist building products for Concrete (EFNARC) which includes specification & guidelines for polymer-modified cementitious flooring applied to a direct finished concrete base or a fine concrete screed.
- H. J. Shearing, *US Pat.* 3778290, 1973 [Hydraulic cement urethane decorative flooring surfaces].

1.3 Polymer Concrete Mortar and Screed Repair and Restoration

Polymer concretes contain reactive polymers which replace cement as binder and are formulated with aggregates, pigments and other solid fillers in the form of mortars and screeds for new construction and for repair. Mortars typically have a rougher texture than screeds because they are formulated with coarser aggregate in addition to fine aggregates and binder for optimum strength – screeds are smoother mixtures formulated without coarser aggregates to create level and fine-grained finishes. Both types are particularly effective with old concrete as reactive polymeric binders not only bond together aggregates, pigments and other solid fillers, they also provide high levels of adhesion to conventional cement-based concretes as well as masonry and other materials of construction.

Unlike the thermoplastic polymer emulsions used to modify cement, the reactive polymers used in polymer concretes are thermosetting resins, and it is the resin selection that determines the ease of mixing and application, the working life, rate of cure, the sensitivity to the environment during cure, the ultimate adhesion and mechanical properties as well as resistance to chemicals and immersion.

1.3.1 Essential Chemistry and Technology

The hardening mechanism of polymer concrete mortars and screeds involves resin crosslinking rather than hydration of cement – this leads to cured compressive strengths significantly higher than those of polymer-modified concretes. The common commercial types of resinous polymer concretes and mortars are based on epoxy functional (epoxy) and polyurethane (PU) resins due to their high load-bearing properties and good all round chemical and corrosion resistance. PUs that are highly filled with cement, when applied at

6 to 9 mm thickness, provide high impact and abrasion resistance, excellent chemical resistance and also high thermal shock resistance, which are required for frequent steam cleaning of areas that require a high standard of hygiene. Vinyl ester (VE) resin screeds are highly durable and provide maximum protection against corrosive chemicals. Methyl methacrylate (MMA) resin floorings are used primarily where fast curing (hardening) is required at freezing temperatures, and also where non-yellowing and non-chalking finishes are required. Figures 1.2–1.5 present qualitative assessments of the cure and cured characteristics of the key formulated resinous polymer concretes and mortars.

Polymer concrete screeds and mortars that are easy to mix by hand and apply by hand, are three-part systems which comprise a solid aggregate, a liquid resin base and a liquid or solid crosslinker (also referred to as activator/hardener/solidifier/curing agent). The choice of crosslinker, usage rate and, in some cases, catalyst addition rate to convert the resin into a hard/infusible thermoset network can markedly affect reactivity as well as mechanical and chemical resistance properties of the resin and polymer composite when cured.

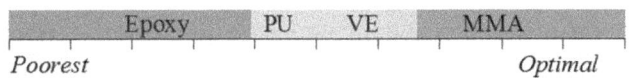

Figure 1.2 Low temperature cure rate comparison of resinous concretes and mortars.

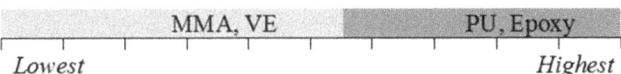

Figure 1.3 Mechanical strength comparison of resinous concretes and mortars.

Figure 1.4 Abrasion/wear resistance comparison of resinous concretes and mortars.

Figure 1.5 Chemical/corrosion resistance comparison of resinous concretes and mortars.

For application at thicknesses of 5–10 mm (200–400 mil) the solid aggregate is typically micronised silica quartz, although granites and bauxites also find use as do coloured quartz sands, where more aesthetically pleasing finishes are required. Washed and kiln-dried aggregates, as well as resin-bound and pigmented aggregates, are widely available in a range of particle sizes to facilitate blending to ensure optimum packing density to preclude voids in applied resin bound compositions.

Fact File: Aggregate Packing Density

Optimisation of aggregate packing density is just as important for enhancing the performance of polymer concrete mortars and screeds as it is for cementitious concretes and mortars. For a simple mix based on a single-sized aggregate in a liquid resin or cement paste binder, the volume of liquid/paste must be larger than the volume of gaps within the aggregate packing in order to fill up all the gaps between the aggregate particles so as to drive away the air voids in the mix (Figure 1.6). By using a combination of aggregates with appropriate size distributions, known as graded aggregate, it is possible for the smaller aggregate particles to fill in the gaps between each of the larger aggregate particles, leading to a smaller volume of gaps (Figure 1.7) within the framework of a well-distributed and well-oriented packed aggregate. This has the added advantage of reducing the volume and cost of expensive liquid resin binders needed to fill up the gaps within the aggregate packing, and minimising polymeric resin binder degradation in use when exposed to weathering and chemicals. The size, distribution, or grading of the aggregate also has an important bearing on the liquid/paste demand and hence the workability of concretes, mortars and screeds – by keeping the volume of liquid/paste binder the same, the use of a graded aggregate increases the volume of binder in excess of that needed to fill up the gaps within the aggregate framework; excess binder aids dispersion of the aggregate particles and provides a coating of resin for each aggregate particle (Figure 1.8) boosting the workability during mixing and application.

The best practical packing density for a three-component system based on dry graded angular aggregate, resin and crosslinking agent gives rise to screed and mortar compositions with pigment or filler volume concentrations of $\pm 74\%$. Pre-wetting of the aggregate, which also has the health advantage of eliminated dusting during mixing, permits higher pigment volume concentrations of 78% to be achieved.

Figure 1.6 Packing density for single-sized aggregates with air voids and increased liquid resin or cement paste binder demand.

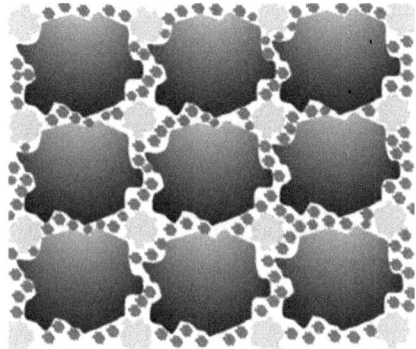

Figure 1.7 Packing density for graded aggregate with reduced air voids and reduced liquid resin/cement paste binder demand.

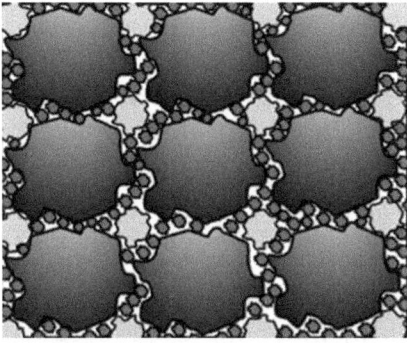

Figure 1.8 Packing density with wetted aggregate providing optimum liquid resin/cement paste binder demand and workability.

Fact File: Pigment/Filler Volume Concentration

Pigment volume concentration (PVC) is a term more commonly associated with the characterisation of paints and protective coatings, being
used to describe how much pigment (and other solid filler) there is in
the paint compared to the amount of binder. As filler to binder ratios
and the pigment volume concentrations increases, many properties of
coating, mortar or screed materials change abruptly – the biggest
change occurring at the critical pigment volume concentration (CPVC),
the point at which there is just sufficient binder to provide a completely
absorbed layer on the pigment surface, as well as all the interstitial
spaces between the pigment particles in a close-packed system, and to
enable bonding to the substrate over which the mix is being applied.
CPVCs are determined practically during the product development
process by varying the filler to binder ratios to achieve optimum properties in physical, mechanical, chemical resistance and ageing testing.
Exceeding the CPVC leads to increased risk of physical damage and
chemical attack.

When deep fill areas are required, pea gravel can be added to bulk
out the mixes. However, gravel can vary – dustier gravels need more
resin binder, smoother gravels need less – so in practice it is best to
add enough freshly mixed polymeric resin binder to wet out the
gravel in a rotary drum mixer (mechanical mixing will be required
to ensure gravel incorporated uniformly) followed by the rest of
the resin binder components and finally the aggregate to complete
the mix.

Two-part systems comprising pre-wetted graded aggregates and a
typically low volume of liquid activator can also achieve the desirable
higher (78%) pigment volume concentrations but require skilled
mixing. Again the right equipment (rotary drum mixer) is required to
ensure a homogeneous blend to avoid soft spots and settling, which
will result in resin-rich weak surface layers.

To ensure successful application of two- and three-part compositions, it is essential that the surfaces they are being applied to
are clean, dry, dust-free and pre-treated with a suitable primer.
Primers, also referred to as conditioners, are required for aged
and new concrete to promote adhesion of overlaid polymeric
repairs, screeds and mortars because they penetrate into the concrete substrate for a better bond and help to eliminate bubbles
and pinholes that can form due to outgassing of concrete. With
highly porous concrete, it is beneficial to use thixotroped primers,

which thin down on mixing and thicken up in position after application.

Application Challenge – Oil Impregnated Concrete

Surface preparation is essential to ensure good adhesion on concrete, even when primers or conditioners are used with overlaid polymeric repairs, screeds, mortars, sealants, coatings and synthetic resin flooring. For optimum adhesion it is essential for any substrate being treated to be clean and dry after mechanical abrasion to remove surface contamination, and for all oil and grease deposits to be removed with a proprietary cleaner/degreaser. SSPC-SP13/NACE No.6 are the standards relevant to surface preparation of concrete. Superficial oil stains can be removed from the surface of concrete by scrubbing with detergents and solvents, application of clay-based absorbent saturated with solvent poultice, or through the use of oil-eating microbes. Where oil has permeated below the surface over time, flame cleaning is the only effective method of treatment. Passing an oxyacetylene blowpipe over the surface to burn away the oil followed by mechanical abrasion to remove the weakened top layer (the flame is hot enough to damage the surface of the concrete) leaves a clean, dry surface which is best primed/conditioned and overlaid immediately.

1.3.2 Fit-for-purpose Reactivity Testing

Working lives of bucket mixes of two- and three-part polymer resin-based concretes, screeds and mortars are shorter than polymer-modified hydraulic concretes due to exotherms associated with crosslinking – heat generated on reaction of the resins accelerates curing in bulk, so in practice freshly mixed buckets are tipped onto polythene sheeting placed adjacent to the repair prior to trowelling/screeding to hold back the cure process until required.

Fact File: Arrhenius Behaviour

MMA, UPR, VE, PU, epoxy and other thermoset resins evolve heat when they react (irreversibly) and cure to form polymers. The heat of reaction varies with the material and the mechanism of cure ranging from 13 kcal mol^{-1} for methyl methacrylate polymerisation, 15 kcal mol^{-1} for styrene/polyester or vinyl ester polymerisation, 24 kcal mol^{-1} for isocyanate crosslinking with a polyol to 47 kcal mol^{-1} for isocyanate foaming with water, and 22 kcal mol^{-1} for thermally catalysed epoxy to

25 kcal mol^{-1} for epoxy crosslinked with a polyamine. All thermoset resins exhibit Arrhenius behaviour in which the kinetics of curing is temperature dependent, which means polymerisation proceeds faster as the temperature is raised. As a rule of thumb, an increase of 10 °C typically leads to a doubling in the rate of a reaction – half the time to cure/harden off and go through to complete reaction, but also half the working life. In contrast, a decrease of 10 °C in ambient temperature leads to a halving in the rate of reaction – doubling the time to cure but not necessarily doubling the working life as solvent-free epoxies and polyurethanes applied to surfaces at near freezing temperatures in-crease in viscosity moving from liquids to solid glass consistency, making them not only unworkable but susceptible to phase separation, which puts an end to any intended reaction (see Chapter 3 for details on sol gel transformations).

Indications of cure/hardening times for polymer modified con-cretes can be obtained from ASTM C191/BS 4550 Vicat needle pene-tration tests where the hard dry time provides an approximation for sufficient strength in the cured repair for pedestrian foot traffic (light loading). Approximation of the time to achieve full cure can be made by determination of the time it takes to achieve ultimate compressive strength according to ASTM D695/ISO 604 by casting multiple blocks or cylinders and crushing samples progressively over time until no further increase in strength measured. Both these laboratory methods are used for comparative or indicative purposes rather than as abso-lute reactivity measurements as environmental conditions (substrate temperature, air temperature, airflow, relative humidity and atmos-pheric carbon dioxide content) all influence the actual cure process on site.

Walk-on times for two- and three-part PU and epoxy resin floor screeds range from 2–6 hours at ± 20 °C (± 70 °F), extending to 6–12 hours at ± 10 °C (± 20 °F) with full cure of screeds and mortars being achieved between 2 and 7 days at ± 20 °C (± 70 °F). The re-activity of vinyl ester and MMA resin screeds can be enhanced relatively easily by increased catalyst loading, so that much faster return to service times can be achieved especially at low ambient temperatures.

Fact File: ASTM C395

The ASTM C395 specification, which relates to chemical-resistant resin mortars for bonding chemical-resistant brick or tiles, includes phenolic and furan resin technologies in addition to epoxy and vinyl esters/polyesters. Specified values for viscosity, chemical resistance, working life, setting time, tensile strength, compressive strength, bond strength, shrinkage and absorption are prescribed – the most critical of which are shared with the requirements for flowable grouts covered later in Section 1.9.

1.3.3   Fit-for-purpose Mechanical and Physical Testing

Adhesion is another important property requiring validation, as unless a polymer concrete, screed or mortar adheres and remains bonded to the substrate to which it is applied, all other properties are irrelevant. ASTM D7234 is the method used to determine adhesion and mode of failure on concrete involving a test apparatus known as portable pull-off adhesion tester, which applies a concentric load and counter-load to the upper surface of a test sample. Samples comprise concrete slabs which are primed and overlaid with the polymer concrete, screed or mortar, which is allowed to achieve full cure before aluminium test dollies are bonded on with a high adhesive strength two-part epoxy glue. After this has fully cured, a cutting tool is used to cut around the edges of the dolly through to the concrete substrate to isolate a specific diameter test area. The test actuator is then attached to the dolly to determine the greatest lifting force (in tension) that a surface area can bear before a plug of material is detached – failure occurring along the weakest plane within the system. It is important to note that measurements are limited by the strength of adhesive bonds between the aluminium dolly, the overlay and the substrate, as well as the cohesive strengths of the adhesive, overlay and concrete substrate. Adhesion values to primed concrete are typically in the range 3–4 MPa (450–600 psi) for polymer concretes, screeds and mortars, and mode of failure is frequently in the concrete as adhesion levels are routinely in excess of the cohesive strength of concrete.

General toughness and resilience to impact from falling tools or other sources of potential mechanical damage is determined correspondingly from the ASTM D256/ISO 180 notched Izod impact test. This measures a material's resistance to impact from a swinging pendulum, and Izod impact is defined as the kinetic energy needed to initiate fracture and continue fracture until the specimen breaks. Standard specimens are typically notched to prevent deformation of the specimen upon impact, and the energy expended in breaking a specimen is recorded in Joules (or foot-pounds) and then divided by the actual dimension in mm (or inches) along the notch of the specimen to give a value in J m^{-1} (or foot-pounds per inch). Polymer concrete, screeds and mortars typically exhibit many times the impact resistance of unmodified concrete, varying between 10 and 110 J m^{-1}.

Tensile strength, which is the maximum amount of stress a material can withstand while being pulled before it starts to be permanently distorted, is often the most critical of all stress forces that a polymer concrete faces in structural applications, but the physical nature of polymer (and hydraulic) concretes makes it difficult to run direct tensile tests.

Flexural strength measurement, however, serves as an indirect tensile test from determination of how much load a non-reinforced concrete beam or slab can bear before it begins to bend. ASTM C580 establishes both the flexural strength and modulus of elasticity in flexure of polymer concretes, screeds, mortars (brick and tile grouts, structural grouts, machinery grouts and >60 mil monolithic sur-facing). Cast or moulded samples of rectangular cross-sections are generally tested in flexure as a simple beam in centre-point loading – the bar rests on two supports and the load is applied by means of a loading nose midway between supports. There are three method options depending on the bound aggregate size: Method A is used for systems containing aggregate less than 0.2 inch (5 mm) in size; Method B is used for systems containing aggregate from 0.2 to 0.4 inch (10 mm) in size; and Method C is used for systems containing aggregate larger than 0.4 inch (10 mm). The resulting values stated in psi units are regarded as standard, and mathematical conversions to SI units are not considered standard. When flexural strength cannot be determined for materials that do not break, tangent modulus of elasticity is determined. Polymer concrete, screeds and mortars typically exhibit flexural strength values between 3600 and 5500 psi (25–40 MPa). Low flexural strength results provide an indication of either excessive flexibility as the material bends and distorts under low force, or that the material is brittle.

Where polymer concretes are used to support structures, it is critically important to determine the compressive strength to validate fitness-for-purpose. The significantly higher strengths of fully cured polymer-bound concretes compared to hydraulic concretes means smaller test specimens are needed in crushing tests in Universal Testing Machines. ASTM D695/ISO 604 stipulate specimens can either be blocks or cylinders – for the ASTM, blocks of dimension 12.7×12.7×25.4 mm

(1/2 by 1/2 by 1 inch) or cylinders which are 12.7 mm (1/2 inch) in diameter and 25.4 mm (1 inch); for ISO, specimens are $10 \times 10 \times 4$ mm for strength and $50 \times 10 \times 4$ mm for modulus. Compressive strength is again calculated from the corrected maximum crushing load and the original specimen cross-sectional area; the modulus from the ratio of stress to corresponding strain based on corrected displacement from the linear plot. The compressive strength values of fully cured polymer concretes, screeds and mortars typically exceed the 20–70 MPa (3000–10 000 psi) range of unmodified (hydraulic) concretes extending up to over 100 MPa (15 000 psi) dependent on workability and filler to binder ratio/pigment and filler volume concentration. The compressive modulus of fully cured polymer concretes can reach 1.3 GPa (185 000 psi), and provides a good measure of rigidity of a repair screed or mortar under sustained compressive forces.

ASTM C267 provides three different methods for screening chemical resistance based on the aggregate particle size of polymer concrete mortars and screeds, through changes in weight and appearance of cast specimens, changes in appearance of the chemical test medium, and from changes in ASTM D695/ISO 604 compressive strength. Immersion testing by these methods is a very effective way of determining whether a polymer concrete mortar or screed repair is suitable for applications involving prolonged immersion, or suitable only for applications involving immersion for short periods/splashing/contact with fumes, or only in environments where accidental splashes or spillages are removed by regular cleaning or, in the case of volatile solvents, by evaporation.

Recommended Reading

- H. H. C. Wong and A. K. H. Kwan, *Packing Density: A Key Concept for Mix Design of High Performance Concrete*, 2005, Proceedings of the Materials Science and Technology in Engineering Conference, HKIE Materials Division, Hong Kong, May, pp. 1–15.
- *EN 1504-3 Products and Systems for the Protection and Repair of Concrete Structures ... Structural and Non-structural Repair*, BSI, 2006, ISBN: 0 580 47658 8.

1.4 Concrete Sealing and Protective Coating

Ready mix concrete generally contains water in excess of that required to hydrate cement to assist in mixing, transportation and workability.

As the concrete hardens and dries surplus water exits, leaving behind a network of fine capillaries and internal pores which readily re-adsorb water coming into contact with the surface by natural capillary action. Water can dissolve chloride salt contamination and transport it inside the concrete and start to corrode the reinforcement. Sulfates present in ground water will also cause great damage.

Even in the absence of salts, it is necessary to seal the surface of concrete to avoid ingress of water; moisture that gets reabsorbed, diffused or permeated inside the concrete can expand and contract under alternating warm and cold temperatures, eventually causing cracking and damage to the mechanical integrity. Furthermore, concrete is alkaline in nature and so inevitably is not resistant to acids – even short-term contact with acid spillages as well as long-term exposure to acids formed from contact with industrial waste or by the dissolution in water of CO_2, SO_2 and NO_x from the atmosphere results in erosion. Acid rain will also lead to corrosion of any steel reinforcement exposed as the concrete erodes and spalls. Sealing also helps make concrete more resistant to salts, grease/oil stains, abrasion, dirt accumulation and prevents it from creating dust.

This section describes general industrial and domestic concrete flooring protection through sealing of the surface – waterproofing, damp-proofing and tanking systems required to protect concrete structures against ground water are described in Section 1.6; concrete bridge deck membrane protection is covered in Section 1.7; protective coatings and linings for concrete tanks and bunds against aggressive chemicals follow in Chapter 1.8; sealing and waterproofing of masonry from weathering is covered in Chapter 2.

1.4.1 Essential Chemistry and Technology

There are two basic approaches with, and categories of, concrete sealers.

Penetrating sealers: used primarily on exterior, non-decorative concrete as a protection from moisture and de-icing chemicals. They are based on combinations of silicates, siliconates, siloxanes and silanes which are supplied in solvent or water and designed to penetrate and react chemically within the pores and capillaries of the concrete to block further moisture penetration. Most types provide invisible protection without changing the surface appearance; some leave a sheen when dry. Most types are breathable, allowing trapped or transient moisture vapour to escape.

Membrane-/film-forming sealers: used to penetrate and bond into the surface pores, and also to form a protective film over the surface of the concrete – they are applied to decorative concrete to provide varying degrees of protection depending upon the chemical type selected. Solvent-borne and water-based acrylics are suitable for both exterior and interior concrete where ultraviolet radiation resistance and non-yellowing are required – these single-part materials are also used where ease of application and economy are important and regular maintenance (recoating) is possible, as they are applied thinly and so wear and mark quickly. For high-traffic areas, high-build solvent- and water-based PUs give thicker films than acrylic sealers, producing highly durable scuff-, staining-, chemical- and abrasion-resistant non-yellowing finishes. Two-part epoxies also provide high-build protective films that are hard, long-wearing and abrasion-resistant. Epoxies are obtainable as clear or pigmented and either solvent-borne, solvent-free or water-based, and, with the exception of the latter, are non-porous so will not allow trapped moisture to escape. Epoxies are not resistant to yellowing.

Two-part polyaspartic systems with high solids contents (as high as 100%), are a type of polyurea with comparable properties to PUs and epoxies but higher resistance to abrasion, faster low temperature cure rates and yellowing resistance due to their UV stability – they can even provide UV protection to underlying polymeric screeds and coatings.

Fact File: Photo-degradation

Some polymers are transparent to ultraviolet (UV) light/radiation and are therefore unaffected; polymers which absorb radiation are degraded in a photo-oxidation reaction which splits chemical bonds, causing cracking, loss of strength/weakening or disintegration. Fading, discolouration and yellowing on exposure to UV is usually the first change seen, although some weathered (exposed externally) polymers form a powdery deposit at the surface in a circumstance known as chalking.

Degradation increases with exposure time and intensity of the UV, and inversely with the wavelength of the radiation so it is the shortest wavelength UV radiation which causes the most polymer degradation and harm. Fortunately, the earth's atmosphere blocks UVC, the shortest of the Sun's UV wavelengths, almost entirely, and most of the UVB. Unfortunately, the fraction of UVB radiation remaining in UV light after passing through the atmosphere is of sufficient intensity to cause damage – attack, however, is heavily dependent on cloud cover and atmospheric conditions, so susceptibility of exposed polymers to photo-degradation varies with the prevailing natural environment.

Some polymers are degraded by water alone through hydrolytic cleavage, and others by temperature alone through thermo-oxidation, and although temperature does not affect the primary photo-chemical reactions initiated on UV exposure, natural weathering involving exposure to UV and water from rainfall or condensation, and elevated ambient temperatures causes acceleration in secondary reactions involving the by-products and enhanced degradation rates.

1.4.2   Fit-for-purpose Testing

Penetrating sealers which fill the pores of concrete do not help exposed aggregate stones form loosening and will not enhance appearance as they are basically invisible when dry. However, film-forming sealers applied to exposed aggregate surfaces can improve both appearance and performance, including protection against dusting, stains, abrasion, freeze-thaw damage and efflorescence. The published standards for membrane-/film-forming liquids as sealers for hardened concrete are ASTM C309 and ASTM C1315, which also cover the requirements for curing compounds for fresh concrete.

Fact File: Curing Compounds for Concrete

Film-forming sealers applied to freshly cast concrete as soon as the concrete is hard enough to walk on, are also known as "cure and seal" materials. They are used to form a film that retains mix water, allowing normal hydration – the film doubling and remaining as a protective surface sealing treatment to reduce dusting and increase stain resistance of the newly cured concrete.

The standards embrace three types of membrane-forming sealers (I – clear; II – clear with fugitive dye; III – white pigmented), all with the fundamental requirement of a maximum permissible moisture loss through a membrane film in the ASTM C309 stipulated test. Penetrating sealers which are chemically reactive inside concrete rather than membrane-forming do not meet the intent of the specification, so other test methods are used for these and the membranes to validate resistance to chloride salt permeability in the classic ASTM C1543 salt ponding test, and the ASTM C1202 rapid chloride permeability test.

The ASTM C309 and C1315 standards also identify three classes of membrane based on resistance to degradation by UV light (A: non-yellowing; B: moderate yellowing; C: severe darkening and yellowing) following exposure to fluorescent UVB, heat, air and water stresses in

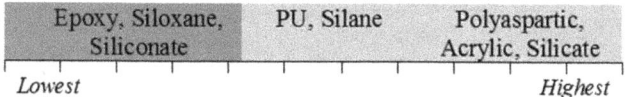

Figure 1.9 UV/oxidation resistance comparison of penetrating sealer and film-forming concrete sealer technologies.

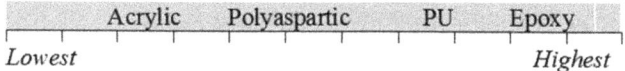

Figure 1.10 Resistance to chemical corrosion comparison of film-forming concrete sealer technologies.

the ASTM G154 test protocol. Figure 1.9 presents a qualitative assessment of the resistance to oxidation and degradation on exposure to UV light for the various penetrating sealer and film-forming sealer polymer technologies.

White-pigmented membrane-forming compounds which serve the additional purpose of reducing the temperature rise in concrete constructions exposed to radiation from the sun are discussed in Chapter 2.

General requirements for film-forming sealers for concrete include good adhesion-promoting qualities as determined in ASTM D4541 coating pull-off strength test, and resistance to change in appearance, blistering, softening, swelling or loss of adhesion in contact with corrosive alkalies and acids as specified in ASTM D1308 – Figure 1.10 presents a qualitative assessment of the resistance for the various film-forming concrete sealer technologies.

Application Challenge – Concrete Outgassing

Two-part polymer systems, particularly those based on epoxy, can be temperature and humidity sensitive so sealers/protective coatings should always be applied to porous concrete at the warmest part of the day when temperatures are decreasing, not increasing. The reason for this is that air inside the pores and capillaries of concrete seeks to escape as the temperature rises and result in outgassing bubbles in any sealer or coating applied. The phenomenon increases with the porosity and presence/number of blow-hole (bug-hole) surface defects resulting from the migration of trapped air (and to a lesser extent trapped moisture) in vertical and overhead surfaces of newly formed concrete. These require filling prior to application of a sealer intended as an impermeable barrier coating. Epoxy resin fairing/smoothing screed coatings are used typically to fill holes and eliminate minor irregularities prior to the application of epoxy impermeable sealers/coatings.

> Fact File: Coating Air Entrapment
>
> Outgassing from concrete should not be confused with coating air en-
> trapment arising from incorporation of excess air during mixing of two-
> pack sealers and coatings. Helical mixer blades driven by slow-medium
> speed motors are the best way of ensuring adequate mixing of protective
> coatings for concrete with the least amount of excess air incorporation –
> high-speed paddle mixing should be avoided. Where air incorporation
> from mixing is the cause of disruption in the coating, the surface of the
> affected area needs to be sanded down, dust removed, solvent wiped and
> an additional coating applied to make it good.

1.4.3   Protective Coating Maintenance and Restoration

There are many different technologies that can be used to renovate
old, dull and stained concrete to make it easier to clean. Nonetheless,
the high-performance, two-part solvent- and water-based and 100%
solids coatings described above are needed in heavy traffic areas such
as warehouses with forklift truck and vehicle usage. It is, however,
possible to use traditional single-part solvent-borne and water-based
coatings in light (pedestrian) traffic areas, and these, typically fast-
drying and easy-to-apply materials are based on long-established hybrid technologies based on acrylic/alkyd, PU/alkyd, epoxy/acrylic and
epoxy/fatty acid ester combinations. Figure 1.11 provides a qualitative
assessment of the ease of use of solvent-borne, water-based and
solvent-free polymeric concrete coating technologies.

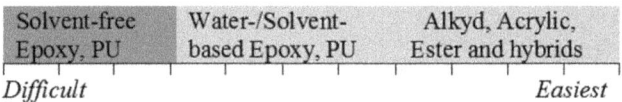

Figure 1.11   Ease of brushing/application comparison of traditional concrete maintenance coating technologies.

When it is necessary to maintain or overhaul traditional and higher
performance concrete coatings it is critically important to remove any
de-bonded, loose or flaking coating residues. Firmly adhering and
undamaged existing floor coatings can then be overcoated, provided
the surface has been cleaned to remove contamination and then
uniformly abraded with fine abrasive pads, and then only with a
compatible maintenance coating type:

- Siliconates, siloxanes and silanes form water-repelling barriers
  inside and just below the concrete surface and require regular re-
  application with similar treatments – they cannot be overcoated/
  overpainted;

- Silicate penetrant sealers react inside concrete to block permanently the pores and capillaries so no maintenance/re-application should be needed;
- Acrylic/PU alkyd and epoxy acrylic/ester hybrid single part paints can only be overcoated with a similar material;
- PU floor coatings can be overcoated with PU, polyaspartic or epoxy;
- Polyaspartic floor coatings should only be recoated with a polyaspartic;
- Epoxy floor coatings can be overcoated with epoxy, PU or polyaspartic.

Where required, addition of ground aggregates into any of the non-penetrative maintenance coating options can be made to introduce anti-slip/anti-skid properties. Fine, medium or coarse particle size quartz silica sands, and even ground recycled polyethylene can be used in problem areas such as walkways, steps, traffic aisles and areas subject to frequent washing where there is a need to reduce the risk of slippages and falls with minimal change to the appearance of the protective coating. Calcined bauxite aggregates, produced by heating high alumina content bauxite and crushing to create sharp edges, are used to provide grip in both dry and wet conditions for anti-skid and for longer-term durability in areas subject to vehicular traffic, such as ramps, warehouse aisles and production wet areas. Non-slip aggregates are applied typically whilst the coating is still wet and active by broadcasting evenly over the surface, with the excess being swept off when the coating is dry. The amount of aggregate used depends on the desired profile; over-application should be avoided as this leaves a surface which is difficult to clean.

Recommended Reading

- *EN 1504-2 Products and Systems for the Protection and Repair of Concrete Structures … Surface Protection Systems*, BSI, 2004, ISBN: 0 580 45057 0.
- *High-Performance Organic Coatings*, ed. A. S. Khanna, Elsevier B.V., 2015, ISBN: 978-1-84569-265-0.

1.5   Synthetic Resin Flooring

Floorings based on uncoated concrete wearing screeds are difficult to clean, especially when exposed to spillages of oils and fats that are easily absorbed and resist routine washing methods. Although penetrating sealers and thin film-forming membrane coatings can be applied to improve hygiene and aesthetic appearance, overlaid synthetic resin floorings are the materials of choice where long-term physical and mechanical integrity retention is needed. Impact resistance in particular is improved by using coatings greater than 3 mm (120 mil) in thickness as they are capable of distributing and dissipating impact forces; abrasion and wear resistance of a floor can also be improved with a synthetic resin flooring overlay based on an appropriate resin binder, aggregate and filler selection.

1.5.1 Essential Chemistry and Technology

EN 8204-6 classifies synthetic resin flooring protection into a total of eight different types:

- Brush, blade/squeegee, roller or spray applied floor seal (type 1), floor coating (type 2) and high build floor coating (type 3) protection systems, which are explained in the preceding Section 1.4 – these are laid down at less than 150 μm (6 mil), 150–300 μm (6–12 mil) and 300–1000 μm (6–40 mil), respectively, and are used as cost-effective solutions for improving the appearance of smooth concrete in areas subject to light pedestrian traffic, and to prevent dusting, staining and attack of concrete from occasional chemical spillage in the absence of mechanical damage.
- Trowel-finished Epoxy, PU and MMA bound resin screed flooring (type 6) and heavy-duty resin flooring (type 8) systems which are described in preceding Section 1.3 – these are applied in excess of 4 mm (160 mil) and 6 mm (240 mil), respectively, and are used where existing concrete floors are not level or damaged, where good chemical spillage resistance is required, and for optimum load-bearing properties in industrial and high-traffic areas.
- Multi-layer flooring (type 4) is a concrete floor protection solution which is a "sandwich" system made by embedding aggregate within multiple layers of type 2 floor coatings or type 5 flow-applied flooring. Type 4 flooring is also commonly known as single or double "broadcast" flooring, and requires a specialist and skilled application technique to obtain an even distribution of aggregate over the surface of a wet epoxy, PU, polyurea or MMA basecoat and intermediate coatings – diligence is also required to ensure thorough removal of excess aggregate when sufficiently cured. Multi-layer synthetic resin flooring is typically applied at thicknesses in excess of 2 mm (80 mil), depending on the

intended use (medium duty – regular foot traffic, frequent fork lift truck traffic, occasional hard plastic-wheeled trolleys; heavy duty – constant fork lift truck traffic, hard plastic wheeled trolleys, some impact). Invariably, broadcast flooring surfaces are sealed with a clear or pigmented solvent-free epoxy, linear PU or an MMA topcoat depending on desired aesthetic effect and slip resistance – all types exhibit good resistance to chemicals upon occasional exposure from spillage.

• Flow applied flooring (type 5) and heavy duty flowable flooring (type 7), are also known as "self-smoothing" or "self-levelling" concrete floorings due to their smooth surface finishes and are the most challenging to formulate, as many factors can adversely disrupt the intended appearance. Applied at 2–3 mm (80–120 mil) and 4–6 mm (160–240 mil), respectively, with the aid of trowel/comb/notched squeegee/pin rake to cover up the concrete sub-floor, all require finishing with a spiked roller to ensure proper adhesion, levelling and to assist in releasing trapped gas. Both types (5 and 7) are made from either solvent-free epoxy or PU resin binders for optimum balance of load bearing, wear resistance, cleanability and resistance to chemical attack.

The high slip potential on smooth surfaces of types 1, 2, 3, 5 and 7 synthetic resin flooring can be reduced with matte or silk grade variants, or with an aggregate scatter, without compromise to the desired decorative requirement.

Fact File: Electrostatic Discharge Control

All *in situ* applied synthetic resin floor seals, coatings and flooring are inherently insulating in nature (electrical resistance $>10^{12}$ Ω), and for many applications, they need to meet the requirements of IEC 61340-1 and ANSI ESD S20.20 standards to control the risk of electrostatic discharge. This is done in conjunction with adhesive-backed copper tape connected to grounding points under conductive primers/floor seals (electrical resistance $<10^5$ Ω) and static-dissipative coatings and flooring (surface resistance $>10^5$ Ω but less than 10^{12} Ω) through which charges flow to ground slower than with conductive materials.

Incorporation of carbon and other electrically conductive fillers permits the formulation of appropriate variants of most types of primers, sealers, coatings and synthetic resin flooring. At a given amount of conductive particles, called the percolation threshold, a continuous network of filler is formed across the matrix and the material undergoes a sudden transition from insulator ($>10^{10}$ Ω) to conductor ($<10^5$ Ω).

Application Challenge – Carbamation of Epoxy Coatings and Flooring

Surface whitening of two-part epoxy coating and synthetic resin flooring materials occurs typically when applied to too-cold substrates where an undesirable side reaction known as carbamation takes place. Certain polyamine-based crosslink agents have limited compatibility with epoxy resins at low ambient temperatures and have the unfortunate tendency to separate, giving rise to tacky surfaces in a phenomenon known as blush or bloom. The more basic polyamines react with atmospheric carbon dioxide in preference to, and in competition with, epoxy resin to form carbamic acids (eqn (1.3)) and amine carbamates (eqn (1.4)). These reversible add-ition products are unstable and revert to amine and carbon dioxide in the presence of water (humidity or excess) generating carbonic acid (H_2CO_3) at the surface, which then reacts irreversibly with exuded polyamine to form undesirable white patches of amine bicarbonate salts (eqn (1.5)).

$$RNH_2 + CO_2 \leftrightarrow RNHCO_2H \qquad\qquad (1.3)$$

$$2RNH_2 + CO_2 \leftrightarrow RHNCOO^- {}^+NRH_3 \qquad\qquad (1.4)$$

$$RNH_2 + H_2CO_3 \rightarrow RH_3N^+ \; HCO_3^- \qquad\qquad (1.5)$$

The amine bicarbonate salts permanently mar the surface appearance of an epoxy coating or flooring as they are insoluble in water or solvents – washing with acids or alkalies does not deal with the inherent stickiness and tack at the surface. Overcoating of the problem surface is not an option either, as adhesion will be problematic. Tenting and heating of cold substrate areas prior to application is one way of avoiding the problem, and critically the use of high carbon dioxide emitting fuel heaters should be avoided to prevent enrichment of the local atmosphere to evade the carbamation issue.

1.5.2   Fit-for-purpose Flow Testing

Validation of the self-levelling nature of flow applied (type 5) and heavy duty flowable (type 7) flooring is made in line with ISO 2431/ ASTM D1200 by measuring the out-flow of fresh mixes from a filled Ford Cup held open for a limited time and fixed at a set height above a glass plate at a set temperature. The flow diameter of the material on the plate is measured after a further set period, and visual exam-ination of the surface and cut-profile can be made to verify air release properties. Flow diameters in the range 15–25 cm (6–10 inch) after a 5 minute release and further 5 minute flow at 20 °C from 160 mm above the plate are generally taken as an acceptable indication of fitness-for-purpose flow for self-smoothing resin flooring.

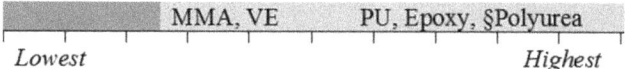

Figure 1.12   Adhesion of synthetic resin flooring technologies § Polyurea
used in Type 4 multilayer flooring.

1.5.3   Fit-for-purpose Adhesion Testing

Bond strength/adhesion testing by ASTM D7234/EN 13892 pull-off adhesion strength measurement and inspection of mode of failure can be used to verify fit-for-purpose surface preparation, primer selection and system mixing/application technique. Bond strengths for epoxy and PU resin flooring to primed concrete are typically greater than the cohesive strength of 25 N mm^{-2} concrete. Figure 1.12 presents a qualitative assessment of the relative adhesion on concrete for the formulated resin flooring types.

Fact File: Concrete Laitance – Removal and Surface Preparation Essentials

Laitance is the weak, powdery layer of cement dust, lime and aggregate fines that may appear on the surface of new concrete that has been over-watered, or allowed to dry prematurely, or even over-trowelled when cast. In order for any resin coating/flooring to bond properly to a new concrete subfloor, all laitance, loose particles and visible dirt or grime needs to be removed to ensure adhesion to sound concrete and not just to a layer of dust on the surface. This is as important as appropriate surface roughening for establishing a good mechanical bond with new concrete.

Rigorous surface preparation of old concrete prior to overlay with a synthetic resin floor is also critical, even when using proprietary two-component epoxy primer systems that are tolerant of residual oil contamination in concrete. It is essential that gross surface contamination is removed by cleaning with an industrial detergent, rinsing with water to avoid any soap film residues, and drying prior to brushing the surface thoroughly with a stiff bristle or wire brush to ensure removal of visible dirt, dust and any loose particles. It is also necessary to vacuum the surfaces thoroughly immediately before applying a resin coating/flooring to ensure a bond to sound concrete.

1.5.4   Fit-for-purpose Mechanical Strength Testing

Methods of tests for *in situ* applied concrete flooring protection systems are described in a variety of national and international

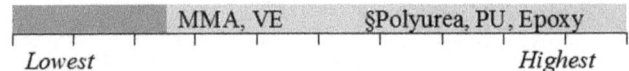

Figure 1.13   Mechanical strength of synthetic resin flooring technologies §
Polyurea used in Type 4 multilayer flooring.

standards with the key mechanical properties required to validate
structural integrity being BS 13892/ASTM C580 flexural strength and
BS 13892/ASTM C873/ISO 604 compressive strength. Figure 1.13
provides a qualitative assessment of the mechanical strength of the
various formulated resin flooring types.

1.5.5   Fit-for-purpose Physical Testing

There are a number of physical properties which are required to
confirm the likely performance in service and "soundness" of synthetic resin flooring. Firstly, impact resistance, which is evaluated in
ISO 6272 rapid-deformation falling weight/BS 13892 drop hammer
tests identifies the minimum mass and/or drop height to cause
cracking or delamination from a substrate. Here, PU, polyurea and
epoxy flooring screeds outperform their MMA counterparts, with PU
inherently most resistant to impact. Figure 1.14 provides a qualitative
assessment of the relative resistance to impact for the formulated
resin flooring types.

Abrasion and wear resistance is evaluated by the various methods
prescribed in ASTM C779 with a variety of equipment to simulate
different types of abrasive action; ASTM D4060 is used to emulate
wear by the dry sliding method. In all cases, polyureas provide the
highest resistance to abrasion/wear loss. Figure 1.15 provides a
qualitative assessment of the abrasion and wear resistance of the
various formulated resin flooring types.

Slip resistance of pedestrian walkways/skid resistance of roadways
can be measured simply in the BS 7976-2/ASTM E303 pendulum test,
which determines the friction between a rubber slider attached to
the foot of a swinging pendulum and a hard, smooth floor surface
under wet or dry conditions. Resulting "pendulum test values" (PTVs)
are used to classify slip potential: a PTV of $0–24 =$ High;
$25–35 =$ Moderate; $36+ =$ Low slip potential. There are numerous
other standard methods and other methods involving proprietary
equipment for determining dynamic and static coefficient of friction
(CoF) values for wet/dry flooring – here, higher CoF values indicate the
lower possibility of slipping, and smaller CoF values a greater danger

MMA, VE
Epoxy
§Polyurea, PU
Lowest
Highest

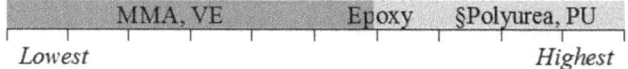

Figure 1.14 Impact resistance of synthetic resin flooring types.

VE, MMA
Epoxy
PU, §Polyurea
Lowest
Highest

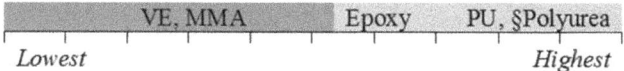

Figure 1.15 Abrasion/wear resistance of synthetic resin flooring types.

Epoxy, MMA, PU, VE
CoF
0.2
0.4
0.6
0.8
1.0
Least
Highest

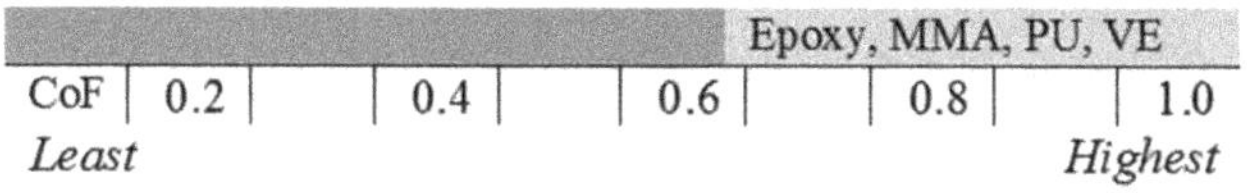

Figure 1.16 Slip/skid resistance of synthetic flooring resin types.

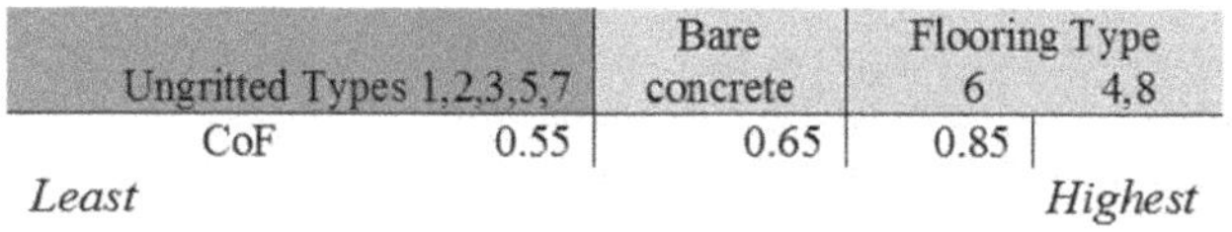

Ungritted Types 1,2,3,5,7
Bare concrete
Flooring Type
6
4,8
CoF
0.55
0.65
0.85
Least
Highest

Figure 1.17 Slip/skid resistance of synthetic resin flooring by EN 8204-6 Type.

of slipping. Generally, any CoF over 0.55 is accepted as safe for foot traffic in conventional shoes. Figure 1.16 is a summary of the relative slip/skid resistance as measured by CoF for the formulated resin flooring technologies.

There is little practical difference in the CoF between the various resin binders used in flooring, so when it comes to the design of a slip resistant flooring it is the selection of type and particle size of abrasion resistant grit/aggregate which dictates performance. With types 1 (floor seal), 2 (floor coating) and 3 (high build floor coating) floor protection systems discussed in Section 1.4, grit added to, or broadcast over the surface, of these smooth finishes as well as types 5 (flow applied flooring) and 7 (heavy duty flowable flooring) resin flooring can be used to increase the CoF for applications which would otherwise become slippery when wet. Types 4 (multi-layer flooring), 6 (resin screed flooring) and 8 (heavy duty resin flooring) are formulated typically with "heavy" grits and highly abrasion resistant aggregates to enhance wear resistance and are as a result slip resistant without additional surface treatment. Figure 1.17 is a summary of the relative slip/skid resistance as measured by CoF for the formulated resin flooring types designated in EN 8204-6.

Recommended Reading

- *BS 8204 Screeds, Bases and In situ Floorings. Synthetic Resin Floorings. Code of Practice*, BSI, 2010, ISBN: 978 0 580 73670 4.

1.6   Waterproofing and Tanking

Concrete structures, such as basements, secondary containment bunds, tunnels and building foundation walls originating below ground level which are subject to rising damp, penetrating damp and water pressure can be protected by sealing the walls "below grade" with waterproofing "tanking" barrier protective coatings or membranes, or in extreme cases where free flowing water cannot be diverted or there is a risk of flooding, by lining with a cavity drain membrane system. These steps are required to prevent foundation failure, distress, and other moisture-related problems.

Fact File: Hydrostatic Pressure

Where there is a hydrostatic head of pressure from water build-up in the ground surrounding a building or structure, waterproofing is necessary to prevent water being pushed through capillaries, tiny cracks or holes in the concrete foundation and walls. Hydrostatic head pressure is greater in terrains that do not allow much water to flow away from a building, such as on clay and solid rock. Waterproofing underground structures is known as "under-/below-grade" application work ("above-grade" waterproofing applications refer to those above ground but below the roof location on a building). Application of waterproofing barrier protection to the internal walls, base slab and where necessary the inside roof to protect the envelope of the structure underground against water ingress, is commonly referred to as "tanking".

1.6.1   Essential Chemistry and Technology

Waterproofing barrier protection applied externally to the "positive" side of underground concrete structures is largely provided by rubberised laminate sheet membranes, PE/PP/EPDM/TPU reinforced thermoplastic sheet compositions, sodium bentonite geotextile matting, or cementitious/bitumen/polymeric coatings. Internal waterproofing to hold back ground water on the "negative" side also features cementitious coatings, bituminous coatings, and polymeric coating membranes and sealants.

Cementitious tanking slurries are routinely polymer modified hydraulically setting cements designed to penetrate into and block the surface capillaries of concrete and repair micro-cracks by forming a monolithic bond to become an integral part of the structure. Polymer modification – used to increase adhesion, flexibility, crack resistance and improve water impermeability – is introduced through acrylic powders pre-mixed in with the cement–aggregate component, or through SBR/SA copolymer emulsions supplied as gauging liquids which also act as bonding agents/primers between coats. The thickness of the cementitious coating required depends on the level of waterproofing required and the external water or hydrostatic pressure to be dealt with. Cementitious tanking slurries can also be applied to cement renders and masonry to make structures waterproof without the need for external membranes – when dry, they can be plastered to provide a smooth surface to decorate where aesthetics need to be considered.

Solvent-borne bitumen or bitumen/rubber latex emulsion paints are used for below-grade waterproofing applications where their susceptibility to "bleed" through any coatings subsequently applied over them necessitates an additional sealing primer. Rubberised asphalt applied in the form of bonded sheet, as a hot-melt coating, as a brush or roller applied solvent-borne paint, or a spray applied liquid emulsion paint is also used widely for below-grade foundation wall external waterproofing.

So-called "high performance" solvent-free, two-part epoxy, PU and polyurea coatings are used for waterproofing concrete water tanks and reservoirs as well as concrete subject to exposure to aggressive chemicals and even traffic.

Silicate-based penetrating sealers which react, harden and block the inside of the capillaries, pores or micro-cracks in concrete are also a proficient waterproofing solution in the same way that crystalline permeability reducing admixtures incorporated when the concrete structures are cast. Once cured, they cannot be pushed out, irrespective of hydrostatic pressure. In emergency situations where it is necessary to stop live leaks occurring in uncoated or rendered concrete due to hydrostatic water flow, and for pressure areas where a wall meets a floor and there is running or weeping water, there are rapid setting cement plugs which can be used with manual compression as an effective repair. Injection of resins to resolve negative-side leaks and waterproofing is covered later in Section 1.11.

1.6.2   Fit-for-purpose Testing

There are no standard requirements which define concrete water-proofing and tanking materials as fit-for-purpose but ASTM D5385 is the standard test method used to determine the hydrostatic pressure resistance of waterproofing membrane sheets, and there are also a number of standard methods including ASTM C1306 and ISO 7031 used to establish the resistance to penetration of waterproofing systems to positive and negative water pressure. Table 1.2 provides an indication of the water pressure resistance capabilities of the various waterproofing and tanking systems.

The permeability/resistance to the passage of water vapour through waterproofing membrane film and sheet is evaluated from water vapour transmission (WVT) rates determined in line with methods described in the ISO 15106 and ASTM E96 standards. Water absorption measurements, attained by a variety of methods cited by waterproofing membrane manufacturers, are also routinely used to validate specific waterproofing system capabilities.

In addition to pull-off adhesion testing to concrete, there are numerous application-specific physical and mechanical tests to ensure fitness-for-purpose for a wide range of applications which include: resistance to puncture; resistance to chisel impact; resistance to aggregate indentation; resistance to thermal shock/heat ageing; crack-bridging; and tensile strength/elongation at break. As such, it is difficult to compare and contrast the performance of the many different waterproofing solutions. However, the environmental impact of the various solutions from an installation perspective is more easily reckonable, as indicated in Figure 1.18.

Table 1.2   Water pressure resistance of waterproofing and tanking systems.[a]

Water pressure	Positive	Negative
Membrane sheet	5–10 bar	
Cementitious coating	12–13 bar	7 bar
Polymeric coating	7 bar	3 bar

[a]Where 5 bar (500 kPa) pressure equals 50 m head of water.

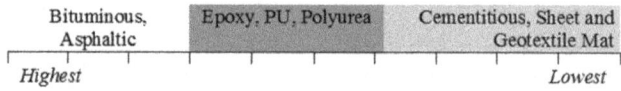

Figure 1.18   Environmental impact during installation of waterproofing and tanking systems.

Recommended Reading

- *BS 8102 Code of Practice for Protection of Below Ground Structures Against Water From the Ground*, 2009, BSI, ISBN: 978 0 580 59399 4.
- *ASTM D7832 Standard Guide for Performance Attributes of Waterproofing Membranes Applied to Below-Grade Walls/Vertical Surfaces (Enclosing Interior Spaces)*, ASTM International, West Conshohocken, PA, 2014.

1.7 Bridge Deck Membranes

Concrete bridge deck membranes are a specific type of waterproofing barrier applied on top of concrete and subsequently protected by another material that functions as a traffic surface.

1.7.1 Essential Chemistry and Technology

Waterproofing membranes for concrete bridge decks are typically used with other components to improve adhesion of the membrane to the deck and also to the protective riding surface. Generically, they are either pre-formed sheet systems or constructed-in-place liquid systems. Sheet waterproofing systems are based on bituminised fabrics, bitumen laminated boards and a variety of polymer combinations with PE, EP, EVA, PVC and butyl or synthetic rubber elastomers. Constructed-in-place systems can be hot applied asphalts, solvented polymer modified bitumens, or resinous cold applied and chemically curing ^{CT}PU, polyurea, ^{CT}epoxy or MMA resin systems (^{CT}modified with coal tar to lower costs and enhance water resistance).

1.7.2 Fit-for-purpose Testing

The same performance requirements and physical properties in terms of permeability, pliability and resilience referred to in the preceding section also apply to concrete bridge deck membrane fit-for-purpose testing. ASTM D6153 lays out the overall specification requirements for the three types of deck waterproofing membrane systems designed for use on bridge decks protected by an asphaltic concrete overlay – Type I cold liquid-applied chemically curing membrane materials; Type II hot applied elastomeric membrane systems; Type III pre-formed sheet membrane systems. ASTM E96 is the specific test method for water permeability screening for determination of water

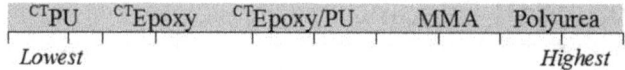

Figure 1.19   Water permeation resistance of constructed-in-place, cold liquid-applied bridge deck membranes, with and without coal tar ^CT^modification.

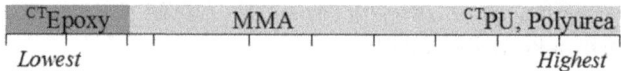

Figure 1.20   Thermal cycling stability of constructed-in-place, cold liquid-applied bridge deck membranes, with and without coal tar ^CT^modification.

vapour transmission rates. For the cold liquid-applied PUs and epoxies, and the spray-applied rapid curing polyureas and MMAs, there are differences in susceptibility to water transmission and absorption. Figure 1.19 provides a qualitative assessment of the water permeation resistance for the cold liquid-applied bridge deck membrane technologies.

Exposure to temperature fluctuations between freezing and high ambient also reveals differences in retention of flexibility, dimensional stability and resistance to physical damage, as summed up in Figure 1.20.

In common with protective coating usage in other applications, waterproofing systems have to provide long-term protection so must maintain adhesion and integrity throughout the service life of the system. ASTM D7234 pull-off testing is used to ensure adhesion of the waterproofing system (membrane and any primers/bonding agents) is sufficient to result in cohesive concrete substrate failure rather than a minimum adhesion value reading.

Application Challenge – Coal Tar Replacement

Coal tar, or pitch, modification is a traditional cost-effective means for enhancing the water resistance and anti-corrosive performance of epoxy and PU paints and coatings as well as waterproofing membranes. Unfortunately, the presence of polycyclic aromatic hydrocarbons (PAHs), in amounts which vary with the origin and type of coal from which they are derived, have long been a cause for concern for health, safety and environmental reasons. Although still used extensively in bridge deck waterproofing membranes, coal tars as modifiers for epoxy and PU have been replaced for paint and coating applications with more expensive, but safer, synthetic light coloured hydrocarbon resins, diluents and plasticisers. Where cost restraints prevail, naturally occurring and crude oil derived thermoplastic bitumens are used as safer alternatives.

Recommended Reading

- *ASTM D6153 Standard Specification for Materials for Bridge Deck Waterproofing Membrane Systems*, ASTM International, West Conshohocken, PA, 2015.
- A. R. Price, *UK Transport and Road Research Laboratory Report No.248*, 1990, ISSN 0266-5247.

1.8 Containment Linings

Coatings are applied as linings to concrete ponds, pools and basins that are constructed to safeguard the subsoil/prevent flooding from excess rainwater run-off or dirty water, and to sewage containment tanks to protect subsoil and groundwater against pollution. Coatings are also applied to concrete tanks designed to contain chemicals and to secondary containment structures/bund walls to protect the environment from spills where chemicals are stored in tanks and pumped through pipes. Containment linings are required to control and resist permeation/chemical attack, and to protect the underlying concrete from corrosion, mechanical damage or loss of structural integrity.

Fact File: Drinking/Potable Water

Concrete tanks and pipes used to store and transport drinking/potable water require lining with a coating that provides protection for the concrete without tainting the contents. There are no universally recognized and accepted international standards or treatments for drinking water, but the European Drinking Water Directive and the Safe Drinking Water Act in the USA compel legal compliance aimed at reducing the risk of exposure to contaminants which are deemed to constitute a danger to human health. Applied coating systems need to be resistant to tainting water in the face of distinct water treatments undertaken by different water authorities around the world. As such, manufacturers are required to seek approvals, listings and multiple certifications for coatings and ingredients – with numerous coating formulation variations being offered in the various market places, and key ingredients being ruled acceptable or unacceptable depending on specific water treatments to be encountered.

1.8.1 Essential Chemistry and Technology

Protective linings for concrete are formulated in fundamentally the same way as linings for steel when resistance to chemical attack/permeation and mechanical damage are required, and also for

resistance to abrasion and erosion where aqueous slurries are present. Concrete coatings which routinely require an appropriate primer, are typically solvent-free, high-build coatings put on by spray,
brush or roller in two or more coats and are ordinarily based on epoxy,
PU, polyurea, VE or MMA resin technologies – they also incorporate
barrier pigments and fillers to enhance permeability resistance, details for which are in the Chapter 3 discussion on corrosion resistant
barrier coatings for metal.

1.8.2   Fit-for-purpose Testing

Epoxies are the best all-round choice from ASTM D4541 pull-off adhesion testing and for mechanical strength under tension as confirmed from ASTM D638 tensile strength testing. Figures 1.21 and
1.22 present qualitative assessments of the adhesion and mechanical
strength for the key containment coating and lining technologies.

Polyureas and PUs are the top choice from ASTM D4060 testing for
resistance to dry abrasion and for contact with liquids carrying entrained solids which is tested according to ASTM G6. Figure 1.23
provides a qualitative assessments of the abrasion resistance for the
key containment coating and lining technologies.

ASTM C868 testing for resistance to corrosive attack by chemicals
under constant immersion confirms epoxies again provide the best all
round performance. Figures 1.24–1.26 present qualitative assessments of the resistance of the key containment coating and lining
technologies to attack by aqueous acids, aqueous alkalies, solvents
and hydrocarbons.

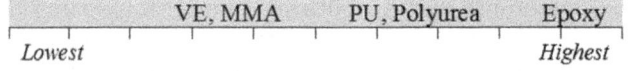

Figure 1.21   Adhesion of containment coating/linings.

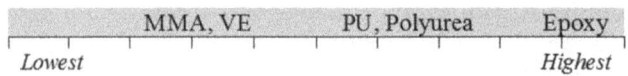

Figure 1.22   Mechanical strength of containment coatings/linings.

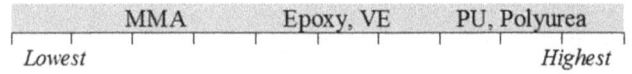

Figure 1.23   Abrasion resistance of containment coatings/linings.

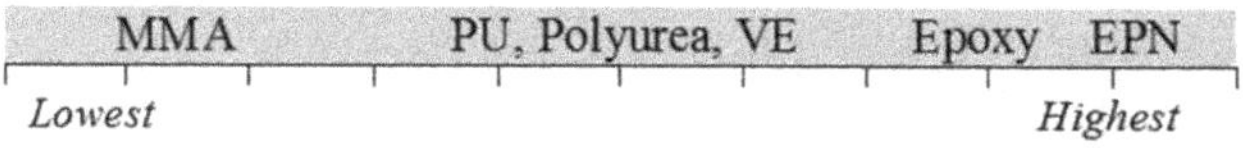

MMA	PU, Polyurea, VE	Epoxy	EPN

Lowest   *Highest*

Figure 1.24   Aqueous acid resistance of containment coating/lining technologies.

PU, Polyurea	MMA	VE	Epoxy

Lowest   *Highest*

Figure 1.25   Aqueous alkali resistance of containment coating/lining technologies.

MMA	PU, Polyurea	VE	Epoxy

Lowest   *Highest*

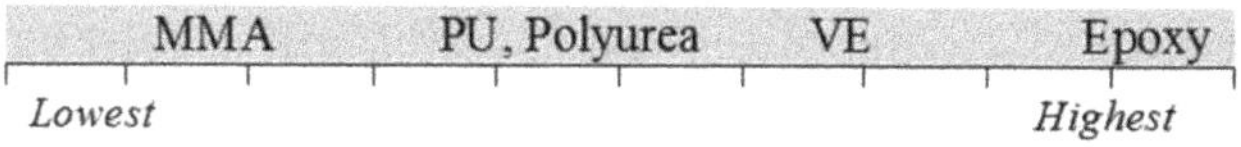

Figure 1.26   Solvent/hydrocarbon resistance of containment coating/lining technologies.

Application Challenge – Bacterial Corrosion of Concrete and Linings

Domestic sewage entering the wastewater system contains organic sulfur compounds which are converted biogenically into sulfuric acid by a combination of sulfur oxidizing bacteria (SOB) and sulfate reducing bacteria (SRB). Coatings and linings used to provide protection for concrete (and steel) from corrosion by sulfuric acid are themselves acutely susceptible to attack in geographical locations with high ambient temperatures and hot seasons – not only are bacterial growth and the biogenic processes more favourable where average temperatures exceed 15 °C/50 °F when biogenic rates double for each incremental 7 °C/12.5 °F rise, so is the severity of corrosion during the warmest periods where water evaporates leading to concentration and increased potency of biogenically produced acids. Unfortunately, traditional waterproofing barrier coatings do not provide sufficient protection as the rate and severity of attack of polymeric coatings increases with concentration and temperature of acid. In these situations, higher inherent temperature resistant and sulfuric acid resistant resin technologies based on solvent-free, high build epoxy novolac resins are employed.

Recommended Reading

- European Drinking Water Directive (DWD), Council Directive 98/83/EC.
- Safe Drinking Water Act (SDWA), 42 USC 300f.

1.9 Flowable Mortars/Grouts

Flowable mortars and grouts are used in concrete construction, repair and joining, filling gaps and voids, for anchoring machinery bolts and as a replacement for steel shims for machinery chocking – they are also used as a backing and reinforcing layer between moving machinery parts, where they provide support for wear liners and are known as crusher backing materials. Their function is structural strengthening, load transfer and damping against impact and shock loads. This section focusses on high-strength load bearing materials – grouts for movement joints are covered in the following Section 1.10; grouts used for the control of water movement or strengthening of soil and rock are outside the scope of this book.

1.9.1 Essential Chemistry and Technology

Formulations are similar to mortars for concrete repair, with pigments and fine fillers such as finely ground silica sand but no aggregates, and with more fluid binders to facilitate application by pouring or pumping into voids, or by spreading with a float or by injection to fill cavities, joints and cracks. The use of more fluid binders can result in filler settlement so significant levels of rheological modifiers are included to keep them evenly dispersed to ensure the homogeneity of the grouts during cure and especially when filling cracks and voids in concrete to ensure greater uniformity of hardened properties and hence monolithic load-bearing behaviour – the modifiers are either viscosity enhancing admixtures which provide structure to the liquid phase, or thixotropes which provide structure to the solid phase. In terms of binder selection, there are two distinct types: cementitious which are Portland cement based with or without polymer modification; resinous which are based on thermosetting epoxy, EPN, UPR, VE or furan resins. Figure 1.27 provides a qualitative assessment of the loading capabilities for cementitious and resinous flowable mortars and grouts.

Portland cement-based grouts are available in sanded, unsanded, pre-mixed, or powdered form and depend on the inclusion of silica

Figure 1.27 Continuous impact loading capability of flowable mortars/ grouts.

sand for their primary strength. Sanded grouts are used for grouting floor and wall joints 125 mil or larger, and unsanded for wall joints less than 125 mil. They are often enhanced with polymers such as acrylic latex blends or with epoxy resin emulsions. Latex modification primarily reduces their absorbency and improves colour retention, whereas the benefits gained from epoxy modification are reduced shrinkage, improved strength and resistance to impact, staining and salt intrusion. Epoxy modified cementitious grouts only offer limited resistance to diluted chemicals and cleaners, so epoxy or furan resin grouts are used especially for installations involving continuous chemical exposure. Modified epoxy grouts are typically three-part systems made up of epoxy emulsion, hardener and pre-blended Portland cement and silica sand.

Epoxy resin grouts, based on 100% solids resins rather than epoxy emulsions, are available as two- or three-part component systems made up of liquid epoxy resin, amine hardener and fillers. They are used for high-strength load bearing applications especially where vibration and impact occur, and where early flexural and compressive strength and dynamic loading are required. Epoxy resin grouts are considered to be chemical resistant although furan resin grouts have superior overall chemical and thermal resistance.

Three-part epoxy novolac resin machinery grouts comprising a 100% solids epoxy novolac resin, amine crosslinker and blended silica aggregate filler are used for resistance to concentrated inorganic acids; three-part vinyl ester grouts based on vinyl ester resins, peroxide catalyst and graded silica aggregate are used for chemical resistance and where fast return to service is required.

Furan resin grouts are available as two-part systems made from furfuryl alcohol polymers and a filler powder containing an acid catalyst used to initiate a thermoset reaction. Filler materials include silica sand, barytes and fibreglass, although for optimum chemical resistance 100% carbon fillers are preferred. Furan grouts are used with brick pavers and quarry tiles where resistance to animal fats, edible oils or citric acid and steam cleaning up to 180 °C/350 °F is required, and also where resistance to a wide range of acids (including hydrofluoric, but not oxidising acids), alkalies and most solvents at constant elevated temperatures is required.

1.9.2　Fit-for-purpose Testing

The relative shrinkage or expansion of cementitious mixtures measured by the ASTM C827 method is critically important for grouts intended to fill cavities or other defined spaces. Otherwise, cementitious grouts intended to provide monolithic mechanical properties with concrete require their load-bearing capacities verified from compressive strength determined according to EN 13892/ASTM C873 and flexural strength measured in line with EN13892/ASTM C78/ISO 178 methods.

There are no specifications or codes of practise for physical and mechanical property values which determine a fit-for-purpose flowable resin grout. Test methods used during development, and where independent test data is required to help define essential suitability involve confirmation that: there is no excessive exothermic reaction when mixed and poured (ASTM D2471) to ensure no cracking on cooling and therefore permit deep pouring; there is good flowability and bearing area achievement (ASTM C1339); and there is no excessive linear shrinkage during setting and curing (ASTM C531). After curing, checks are required to ensure: there is sufficiently high compressive strength (ASTM C579); there is high resistance to creep under a sustained load (ASTM C1181); and the resistance to moisture is low (ASTM C413). Other testing which helps classify resin grouts includes their operating temperature capability (ASTM D648/ISO 75), and resistance to attack by chemicals/water on immersion (ASTM D570).

Application Challenge – Underwater Concrete Repair and Maintenance

Concrete is widely used in the construction of offshore platforms, bridges, harbour walls, coastal defences, canals, dams and other structures where the foundations are positioned underwater. Maintenance of damage due to steel reinforcing bar corrosion, scaling, cracking, spalling or scouring is vital but not as straightforward below the waterline as it is above the waterline, since repair materials need to be compatible with underwater placement techniques as well as be capable of bonding to wet surfaces. Fortunately, both cementitious and resin-based flowable mortars and grouts adapt well to underwater concrete repair and maintenance applications. Cement cures in the presence of water, and hardening of cementitious flowable mortars and grouts underwater is improved by incorporation of pozzolanic materials which are natural or synthetic reactive silicates or alumino-silicates which combine with calcium hydroxide to form insoluble calcium silicate hydrate and calcium aluminate hydrates possessing cementitious properties. Epoxy resin grouts modified with various proprietary moisture tolerance enhancers to ensure cure underwater, are also able to displace water from the bonding surface during application and provide long-term repair and maintenance solutions.

Recommended Reading

- D. M. Harrison, *The Grouting Handbook, A Step-by-Step Guide for Foundation Design and Machinery Installation*, Elsevier Press, 2013, ISBN: 978-0-12-416585-4.
- *ASTM C1107 Standard Specification for Packaged Dry, Hydraulic Cement Grout (Non-shrink)*, ASTM International, West Conshohocken, PA, 2014.
- *ASTM C881 Standard Specification for Epoxy-Resin-Base Bonding Systems for Concrete*, ASTM International, West Conshohocken, PA, 2014.

1.10 Movement Joints

Grouts applied between movement joints are sometimes referred to as sealants, and are used primarily to allow concrete slabs, block walls and pipelines to expand and contract as the ambient temperature changes – they are also used to facilitate movement during ground settlement and seismic tremors, to absorb other sources of vibration, and to act as a sealant to prevent water from getting inside saw-cut slab joints and underneath floor joints. They differ from caulks, which are bulking materials used as a physical non-bonded barrier that prevents the passage of moisture, air, dust, smoke, fire, heat and cold between joints, gaps or cracks in structures.

1.10.1 Essential Chemistry and Technology

Grouts or sealants for movement joints are required to develop a strong adhesive bond to the sides of the concrete slabs they join, and are used in conjunction with backer-rods to create an even depth and support for the grout to form the correct shape (width: depth 2 : 1 ratio for façade joints; 1 : 1 for trafficked/hydrostatic joints) and prevent adhesion at the bottom of the joint. Backer-rods, made from flexible and compressible closed, open or soft-cell foamed polyolefins, are selected with a diameter appropriately larger than the nominal width of the joint easy to develop a tight seal at time of application, and to accommodate the joint gap opening when the ambient temperature falls as the expansion joint system needs to expand to follow the joint movement.

Horizontal joints with wide gaps accept "pourable" self-levelling grouting compositions; narrow horizontal joints and vertical joints require more structured/thixotroped "gun grade" mastics. Movement joint grouts and sealants are also formulated with pigments and fine fillers such as finely ground silica sand but not aggregates, and the common binder types in use are silicone, silyl modified polymers, polysulfide, PU and PU modified epoxies which are all elastomeric in nature (have low Young's modulus and high failure strain) to allow movement in tension, compression and flexure (shear) of up to 50% of the width of a joint. There are a number of other proprietary approaches to expansion joint sealants such as polyurea, plasticized epoxy and high-solids fluoroelastomer with their own niche applications.

Silicones are one component, ready-to-use silicone polymer/ crosslinker blends which on contact with atmospheric humidity cure to form silicone rubbers with elastic properties which remain practically constant between sub-zero and elevated temperatures and on exposure to weathering for extended periods of time. Primer pre-treatments are required on porous substrates. One-part silyl terminated polyethers and polyurethanes both cure on exposure to moisture too, through hydrolysis of the silyl ether functionality and chain extension/cross-linking to form polymeric siloxane linked elastomers with enhanced aesthetics and durability. Pourable grades are not an option with moisture activated technologies.

One-part polysulfide joint sealants cure by absorption of atmospheric moisture, whereas two-component thiol terminated liquid polysulfide polymers require mixing with disulfide bond promoting curing agent component before use – both cure to form flexible and durable rubbers for joints where above average repeated movement is encountered and where resistance to chemical attack from solvents in particular is required.

PU permits the formulation of one- or two-part, gun grade or pouring grade systems with the widest range of performance options for sealing movement/expansion joints. All types show good adhesion to porous surfaces and provide effective resistance to abrasion for use in highly trafficked areas. Two-part epoxies can be formulated to accommodate considerable movement in compression but only limited movement in extension, but PU hybridisation is used to achieve the desired level of movement in tension, flexure and compression – PU/epoxy hybrids offer good resistance to traffic loads and enhanced chemical resistance.

1.10.2   Fit-for-purpose Testing

Grouts used to seal movement joints must adhere well to the surfaces that are being jointed, and must remain bonded when subsequently subjected to tensile, flexural (shear) or compressive forces. Adhesion is therefore the key factor in determining the overall performance of joint seals where there is significant movement; strong adhesion to concrete may not be desirable if the tensile adhesive strength is stronger than the cohesive strength of the sealant to avoid tearing apart during expansion or contraction – another reason why a bond-breaker or release material is used at the bottom of a joint to avoid 3-sided adhesion.

Sealants are subjected generally to lateral and longitudinal shear stresses during use, and ASTM C961 lap shear strength provides a measure of the cohesive strength when subjected to lateral shear stresses as well as providing information regarding the adhesive bond. ASTM C794 is a useful laboratory procedure for determining the adhesion-in-peel strength and also failure mode of sealant/primer combinations to verify whether the failure mode is primarily adhesive or cohesive. As indicated before, cohesive failure is not necessarily better than an adhesive failure providing the adhesive value is sufficient for the application.

ASTM D412/ISO 37 provide the means for validating other important mechanical properties of elastomeric sealants under longitudinal stresses, specifically tensile strength at break, elongation at break and modulus of elasticity. Tensile strength in a sealant is needed to avoid transfer of stress between substrates and cohesive failure and for some applications strength may be more important than elasticity – for example, low-strength/low-tensile modulus may be critical where a sealant is used to join one or more weak surfaces. Low-to-medium-modulus sealants are capable of accommodating significant movement without transfer of stress between joints or stressing the joint sealant. High modulus sealants are more rigid and less able to accommodate movement but are better able to withstand repeated vibration, abrasion and frequent cleaning better than the softer more flexible low modulus ones.

The effects of compression and fatigue are two further important criteria to be considered when validating performance – in practice this is achieved through the ASTM C719 method for adhesion and cohesion of elastomeric joint sealants subjected to water immersion, cyclic compression and expansion movement, and temperature change. This test also helps validate the ISO 11600/ASTM C920 classifications for movement accommodation factor (MAF), which is the total movement range between the maximum compression and the maximum extension that the sealant will accommodate, expressed as a percentage of the minimum joint width – a sealant with a 25% MAF in a 20 mm/787 mil wide joint will have a total movement capacity of 5 mm/197 mil. Typically, movement floor joint sealants have MAF values 5, 7.5, 12.5, 20 or 25%; MAF classifications continue on from 35, to 50, to 100/50% (100% expansion/50% compression) for specialist engineered joints.

Movement joint grouts frequently need to meet many additional requirements specific to the environments in which they are applied. Where exposed to scuffing and mechanical wear they need to be resilient with sufficient hardness and good abrasion, puncture and tear resistance – the hardness of a cured joint sealant is readily determined from its resistance to permanent indentation by a durometer in the ASTM D2240/ISO 868 test protocol, and tear propagation resistance of "nicked" samples is evaluated by ISO 34-2/ASTM D624. Of all the cold applied flexible sealant grouts, PUs provide the highest abrasion and tear resistance.

Where contact with chemicals occurs, either short term exposure to cleaning fluids on floors and spills/leaks on floors or in bunds, or long term exposure in storage tanks, reservoirs or pipelines, testing to determine the effects on physical/mechanical properties of total immersion following the ISO 175 method is used – this verifies resistance to degradation or impermeability especially for joint-sealing systems needed in storage/filling/handling facilities for water-polluting liquids. Figures 1.28–1.30 present qualitative assessments of the resistance of movement joint technologies to attack by aqueous acids, aqueous alkalies, solvents and hydrocarbons.

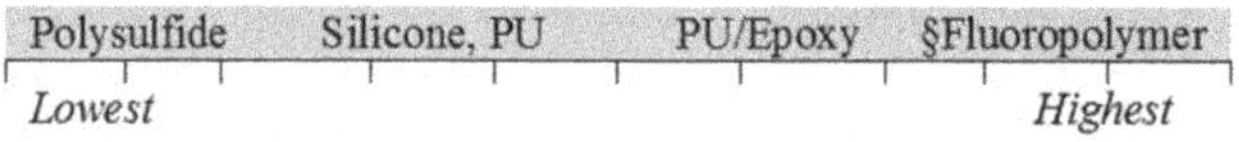

Polysulfide	Silicone, PU	PU/Epoxy	§Fluoropolymer
Lowest			*Highest*

Figure 1.28 Aqueous acid resistance of movement joint technologies.

§Fluoropolymer	PU	PU/Epoxy, Polysulfide	Silicone
Lowest			*Highest*

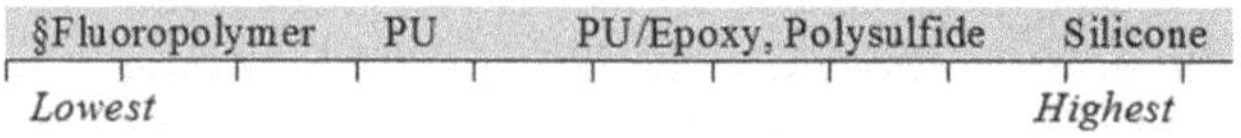

Figure 1.29 Aqueous alkali resistance of movement joint technologies.

§Fluoropolymer, Silicone	PU PU/Epoxy		Polysulfide
Lowest			*Highest*

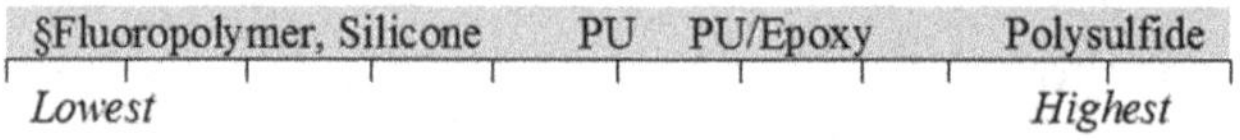

Figure 1.30 Solvent/hydrocarbon resistance of movement joint technologies. § Liquid applied fluoropolymer sealants, not pre-cured fluoropolymer elastomer sheet.

Application Challenge – High Movement Joints in Contact with Aggressive Chemicals

Protecting joints in concrete containment tanks and bund walls subject to high movement and actual or potential long-term contact with aggressive chemicals requires engineered polymer composite solutions. Low modulus elastomeric sealants capable of accommodating high movement do not have the best resistance to the most aggressive chemicals being either readily permeable or quickly losing their physical strength and adhesion. In practice therefore, a multi-system approach is adopted involving packing of the joint with a backer-rod, filling with a sealant grout capable of accommodating the anticipated joint movement, and overlaying with a protective seal or lining of waterproofing/chemical resistant material. The seal can be applied in the form of a strip or complete lining of pre-cured sheet which is bonded to the side of each structural joint using a two component epoxy adhesive – suitable preformed sheets include those used for conventional waterproofing as discussed in Section 1.6, in addition to fluoropolymer rubbers where extreme chemical resistance is required. As an alternative to pre-formed sealant strips or sheet linings, reinforced liquid applied coatings based on epoxy or EPN resins modified with polysulfide elastomers provide resistance to the severest immersion conditions over large surfaces areas with multiple movement joints as well as where there are moving cracks in concrete.

Recommended Reading

- *ISO 11600 Building construction – Jointing Products – Classification and Requirements for Sealants*, International Standards Organisation, 2002.
- *ASTM C920 Standard Specification for Elastomeric Joint Sealants*, ASTM International, West Conshohocken, PA, 2014.

1.11 Injection

Low viscosity resins or grouts are injected into cracks, fissures or voids in concrete to prevent the flow of water or gas out of or into a structure, to provide a flexible repair where further movement is expected, or to make a monolithic repair where structural strength and loading capability need to be reinstated.

1.11.1 Essential Chemistry and Technology

Unfilled resins are used to penetrate fissures and cracks up to 150–200 mil (4–5 mm) wide; fine filler based grouts are used to fill voids and cracks wider than 4–5 mm – both are founded on low viscosity two-part reactive resins to facilitate flow and penetration, and either incorporate thixotropes to permit flow during injection and thickening when dispensing pressure decreases, or function by foaming/gelation to stop flow where needed. The physical nature of the resin/grout as well as location, size and depth of the repair determines the injection method which can be high-pressure compressed air driven pumping or low pressure hand operated pistol for vertical and horizontal repairs, or gravity feed for horizontal repairs. Injection resins and grouts can be structural and flexible, and are based typically on PU, epoxy or acrylic resins as well as micronised cement slurries.

Two-part unfilled PU resins injected into cracks in concrete will react with moisture, and even moving water, to form expanding foam barriers which fill fissures and voids and resist water and gas permeation. Rigid closed cell foaming PU resins are used for static repairs, and flexible closed cell foaming PU resins are used where there is continuing limited movement – reactivity is adjustable on-site by varying accelerator incorporation. In the absence of water in the structure, water can be introduced into the resin component immediately prior to mixing with the activator component and subsequent injection.

Two-part epoxy resins are used where it is necessary to re-establish structural integrity in cracks, fissures and voids by structurally bonding the concrete back together. Very low viscosity resins which are 100% solids and contain no solvents or fillers and with retarded cure rates are used for penetrating the finest of hairline cracks. Very low viscosity but thixotropic liquid resins are used for cracks wider than hairline, and thixotropic grouts are used for the widest cracks, gaps and voids. Epoxy resins and grouts permit repairs inside dry and damp concrete, but are not suitable for live water leaks.

Acrylic injection resin systems are used to seal the finest cracks in concrete in situations where fluctuating groundwater levels occur. They crosslink rapidly to form flexible but solid gels, the volume of which increases or decreases reversibly in wet and dry conditions. The rate of gelation of methacrylate functional acrylic resin systems is controlled by the level of catalyst incorporated into the "A" resin component prior to plural component spray with the "B" component formed by mixing an initiator dissolved in water. Modifiers based on proprietary polyacrylate polymers are included to flexibilise the gels as their volume fluctuates.

Two component grouts based on micronised cement with plasticisers and corrosion inhibitors are used for deep penetration of narrow cracks in concrete, filling voids and cementitious sealing of cracks by injection.

All of the above technologies require injection ports or packers bonded into place, and cracks in vertical surfaces to be sealed, to prevent loss of resin or grout during the injection process – for this, high strength two component thixotroped epoxy or polyester adhesives and repair mortars are used. On completion of the injection repair, the injection ports are broken off and the cured repair surface smoothed off by grinding.

1.11.2   Fit-for-purpose Testing

Given the widely varying technological approaches adopted to resin grout injection, and the fact that injected materials physically block and get locked into cracks, fissures and voids in concrete, there is only a limited amount of testing which defines the fit-for-purpose of a resin grout injection material. EN 12637 deals with compatibility of injection systems with concrete, and even though an injection material when cured gets physically trapped inside the concrete repair, adhesion testing in accord with EN 12618 seems obligatory to confirm that failure in concrete occurs rather than bond failure. EN 12617 defines a method to determine shrinkage of polymeric crack injection materials but the suitability as structural repair materials ultimately calls for compressive strength (various method standards depending on the nature of the injection material: ASTM C109; ASTM C39; EN 12190) and tensile strength (again various methods/ standards: ASTM D412; ASTM D790; ASTM D3574; ISO 527).

Application Challenge – Moving and Active Crack Bridging

Non-moving, dormant cracks in concrete can be filled and the surface sealed with either rigid or flexible repair materials. Active cracks however require flexible repair materials to allow for movement, even when used in combination with mechanical stitching and stapling techniques, to prevent further crack growth. EN 1504 is the standard which defines the performance characteristics required of surface impregnation and coating systems used for both structural and non-structural repair, and also for sealing cracks in concrete to stop leaks, or prevent ingress of water and other damaging chemicals. Many different flexible polymer technologies are used to seal and protect cracked surfaces: they range from polymer modified cements, to single-part liquid applied waterproofing membranes, to two-part reactive flexibilised epoxy/EPN, PU, polyurea, MMA and VE resin based coatings with or without stress relieving reinforcement or pre-cured flexible membrane sheet; the selection depends on the overall function and environment of the concrete. Irrespective of the level of atmospheric or chemical attack to be endured, it is the movement capability of the flexible coating or system applied over the repair which is usually the defining property, and this is validated in the ASTM C1305 or BS EN 1062-7 standard test methods for crack bridging.

Recommended Reading

- *BS EN 12637 Products and Systems for the Protection and Repair of Concrete Structures. Test methods. Compatibility of Injection Products. Compatibility with Concrete*, BSI, 2004, ISBN: 0 580 44411 2.
- *ASTM C881 Standard Specification for Epoxy-Resin-Base Bonding Systems for Concrete*, ASTM International, West Conshohocken, PA, 2014.
- *Concrete Repair Manual*, American Concrete Institute, 4th, 2013, ISBN-13: 978-0870318054.

2 Masonry and Wood Protection and Repair

2.1 Introduction to the Need for Protection and Repair for Masonry and Wood

In addition to concrete, there are various other materials that are used in the building and construction of masonry walls, some of which are more susceptible than others to ground damp, water leakage, driving rain damage and condensation. Untreated stone, ceramic brick and concrete block walls bound together with cementitious mortar are able to absorb water, which over time can lead to dirt pick-up and colonisation by algae, fungi, lichen or moss. Persistent wetting and the presence of hairline cracks in masonry can also give rise to efflorescence and freeze-thaw damage when exposed to frost. Fortunately, polymers can be used in a number of ways to provide straightforward protection and for remediation where problems develop.

> **Application Challenge: Efflorescence in Concrete and Masonry**
>
> Concrete foundations, block and cement based mortars all contain calcium hydroxide (lime) which is formed in the hydration reaction of Portland cement, and they also contain hydroxides and sulfates of sodium and potassium present in cement, aggregates or admixtures. These inorganic materials which are all soluble in water can migrate/get transported as a solution in water through the capillaries and hairline cracks to the surface, where they combine with carbon dioxide in the air to form carbonate salts, which are insoluble in water and deposit on the surface in a process known as efflorescence. In the case of calcium hydroxide, the transportation process is also known as lime leaching. The reaction with

Industrial Polymer Applications: Essential Chemistry and Technology
By William R. Ashcroft
© William R. Ashcroft 2017
Published by the Royal Society of Chemistry, www.rsc.org

atmospheric carbon dioxide to form calcium carbonate is known as carbonation, a process which is deleterious to concrete block as it reduces the alkalinity and leads to corrosion. Most concrete is grey or white so the residue is not always obvious, but on ceramic bricks the white powdery residue is more noticeable.

The problem appears in mortars very quickly after exposure to moist or cool conditions, and can usually be removed by using a stiff brush with water. Efflorescence will occur more gradually through concrete block and brick, and the resulting powdery salts, or tightly bonded crystalline growth arising from repeated wetting and drying cycles, can sometimes be removed by hand-scrubbing or pressure washing. In severe cases, chemical treatment with hydrochloric acid (muriatic acid) is required to wash away efflorescence, but this type of treatment, along with the use of power tools is avoided generally due to the structural damage they can cause as they take off layers of the concrete block and brick with them.

Whilst complete prevention is not possible, controlling efflorescence is achievable by preventing moisture ingress so inherent salts cannot be dissolved and migrate to the surface where the water will evaporate. With freshly cast concrete, water repellent/efflorescence control admixtures can be introduced to make concrete pores hydrophobic and reduce water and salt solution transport. For brick, concrete block and stone, the introduction of a silicone barrier in the capillary channels with a silane or siloxane based exterior surface sealer prevents water penetration and migration.

Wood and wood-based products used in the construction of buildings, where there is a risk of wetting or condensation such as internal wall and roof frame timbers, are generally pressure impregnated with preservatives to resist fungal decay and insect attack. External load bearing timbers, cladding, boards and joinery which are exposed to frequent wetting need to be waterproofed either with oils that penetrate deep inside the wood, or with the same surface membrane-/film-forming (concrete) maintenance coating paint types already discussed in Section 1.4. Traditional flat roofs enclosed with plywood or timber boards which have been waterproofed with layers of roofing felt and tar are particularly prone to localised damage in the form of cracks or splits at the edges as well as blisters and cracks in low spots where water ponds. These and other types of roof designs all need emergency repairs with the liquid applied polymeric roofing systems, which can prove long-lasting if undertaken correctly.

Timber and plywood that is permanently in contact or exposed to wetting with ground, fresh and salt water requires waterproofing and protection from colonisation by slimes, plants and fouling animals. Fortunately, polymeric coatings and resins can be used to seal against the ingress of moisture and provide protection from micro-fouling and macro-fouling organisms. Liquid epoxy resins, recognised initially from their utility as structural adhesives for wooden boats, and then through the pioneering WEST SYSTEM® Wood Epoxy Saturation Technique, are now used widely in primers, undercoats and fillers for the repair, restoration and protection of wood in the marine industry along with other liquid applied polymer types, which are used above and below the waterline for aesthetic reasons as well as foul release or anti-fouling protection.

Recommended Reading

- M. Merrigan, *Masonry Soc. J.*, 1986, 5(1), 1–4.
- *BS EN 335 Durability of Wood and Wood-based Products. Use Classes: Definitions, Application to Solid Wood and Wood-based Products,* BSI, 2013, ISBN: 978 0 580 75630 6.
- *BS 6229 Flat Roofs With Continuously Supported Coverings. Code of Practice,* BSI, 2003, ISBN: 0 580 41305 5.
- M. Gougeon, *Wood and WEST SYSTEM® Materials,* McKay Press, Inc., Midland, Michigan, 5th, 2005 ISBN: 1-878207-50-4.

2.2 Masonry Wall Waterproofing

The application of a skin or protective layer on masonry walls to redirect water or wind and control run-off to prevent infiltration of weather elements is known as cladding—a category of waterproofing solution which prevents the penetration of water in its liquid state and is breathable to allow trapped or transient moisture vapour to escape.

As with controlling the porosity of the surface of freshly cast concrete (see Section 1.4), there are two basic approaches to making porous masonry walls waterproof or water-resistant—invisible penetrating cladding, and protective membrane cladding. The choice of approach depends on aesthetic considerations, as well as the type of wall surface and whether it has been coated before with a paint, render or stucco. Where protection is required without changing the natural appearance of brick or stone on, for example, buildings with protected heritage status, then penetrating cladding is used. Masonry walls made from brick, concrete block or stone, which are either plain or finished with render or stucco, are more commonly protected from the elements with a decorative protective high-build membrane cladding.

Fact File: Render and Stucco

Render and Stucco are terms used interchangeably for external wall mortars made from sand, water and either lime or Portland cement as binder, often with additive fibers and acrylic polymer modifiers to improve structural properties.

2.2.1 Essential Chemistry and Technology

Penetrating cladding: silicates are used on concrete block as they chemically react with free lime in the pores to form tightly bound calcium silicate hydrate which prevents water ingress and restricts water vapour permeability; silane and siloxane functional silicone resin micro-emulsions are used on concrete and other masonry surfaces as they form a hydrophobic three-dimensional silicone resin network within the pores and at the surface to block water ingress and have significant water vapour permeability. Most types provide invisible protection without changing the appearance of the surface, but some leave a sheen when dry.

Membrane-/film-forming cladding: traditional low solids alkyd and acrylic paints used for decorative purposes on masonry tend to deteriorate quickly in high rainfall areas as they are impermeable and trap moisture. High solids/high-build acrylic latex coatings or claddings, however, applied in one or two coats to a total dry film thickness of 10–15 mil (~ 250–400 µm), are formulated with significantly higher pigment volume concentrations to ensure appropriate porosity for water vapour diffusion. High solids acrylics are typically modified with rubber elastomers to reduce their drying time and make them tough and flexible so they are able to bridge any masonry cracks that open and close underneath them. Elastomeric co-polymers of styrene and butadiene (SBRs) are also used as the base for emulsion polymers for masonry coatings as they are fundamentally faster drying than acrylics and also oxidise and chalk quite readily, making them naturally self-cleaning on vertical surfaces in areas with regular rainfall.

Fact File: Self-cleaning Membrane Coatings

Chalking of SBR masonry claddings resulting from exposure to UV is caused by the destructive oxidation of the polymer at the exposed surface. The products of polymer oxidation get washed away when it rains, leaving the pigments and fillers exposed in a loosely bound layer at the surface, which can be easily rubbed off transferring a chalk mark to a fabric or fingertip—hence the term **chalking**. By careful selection of pigment and filler grades, and their loadings, the rate at which weathering occurs can be adjusted to optimise the life of the cladding to a minimum of 10 years depending on the amount of UV exposure and rainfall experienced. In general, the life expectancy of pure and hybrid acrylic claddings, against all types of extreme weather conditions, is up to 15 years.

Silicones are used as cladding binders in their own right, and also as modifiers for acrylics in the formulation of hydrophobic masonry wall coatings with excellent vapour diffusion, resistance to UV degradation and which are also inherently self-cleaning as water and dirt are unable to cling on vertical surfaces—every time it rains the vertical façade simply washes clean.

There are also proprietary super-hydrophobic masonry coatings that mimic the lotus effect due to nanoscopic surface profiles, which significantly reduce the contact area and the adhesion force between surface, dirt and water droplets—this makes them self-cleaning. Furthermore, they do not need algicidal or fungicidal additives, which are used widely in other types of cladding to protect from staining due to algal, fungal and mould growth in-can or in-film. Silicate-based paints made with water glass silicate binders can be formulated as breathable coatings for masonry, and due to their natural alkalinity they provide resistance to algal and mould growth.

Perlite (volcanic glass), ceramic beads and hollow glass microspheres are routinely incorporated into masonry membrane coatings to facilitate application and to enhance insulation performance. Not only do beads and spheres move over each other well and facilitate flow during application (speed and ease) of the cladding, they imbed easily into masonry surface imperfections, and they ultimately also form tightly packed layers and smooth finishes, which provide insulation by reflection, refraction and blocking of heat radiation losses or gains. Energy savings achieved from the layer of vacuum-filled microspheres in the cladding are additional to the fundamental energy savings achieved by simply protecting masonry from water ingress as wet walls are acknowledged to transfer heat twice as quickly as dry walls. Textured surfaces are obtained by incorporation of silica sand and light-weight perlite aggregates when required to mask the imperfections of the masonry or replicate a render/stucco finish—perlite popcorn aggregate is often incorporated where pebbledash finishes are required.

2.2.2 Fit-for-purpose Testing

The testing requirement for penetrating claddings is similar to that described for concrete penetrating sealants in Section 1.4. There are a number of different key physical, mechanical and ageing tests used to confirm the suitability and likely life expectancy of membrane-/film-forming masonry claddings.

Water vapour permeability, or breathability, is determined from water vapour transmission rates by the ASTM E96 and ISO 7783 wet-cup methods prescribed for high relative humidity (93% to 50%) conditions of use—the dry-cup method is reserved for lower relative humidity (50% to 3%) service coatings. Waterproofing against rain penetration is confirmed by evaluation of liquid–water transmission rates in accord with EN 1062 and ASTM E514, and the imperme-ability against wind-driven rain over time is verified in the ASTM D6904 simulation procedure.

Waterproofing membranes need to bridge existing and potential masonry cracks and have sufficient elastic behaviour to withstand movement induced by night–day and seasonal temperature changes, so crack bridge cycling capabilities are validated by the ASTM C1305 and EN1062-7 methods.

ASTM G154 accelerated weathering testing, involving cycles of exposure to fluorescent UV, heat, air and water stresses, is used to evaluate comparative weathering performance properties of membranes by examination of the relative resistance to chalking (ASTM D4214), checking or fine cracking in the surface of the ex-posed membrane (ASTM D660), and fading of any pigmentation (ASTM D1729). Unfortunately, it is not possible to quantify a likely actual service life as UV intensity and exposure times, temperatures, wind and water contact times will vary, even at any given geo-graphical location. Figure 2.1 provides a qualitative assessment of the UV/oxidative degradation resistance of the various waterproofing types.

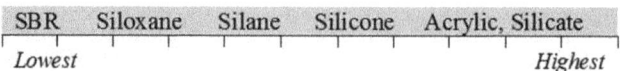

Figure 2.1   UV/oxidation resistance of masonry wall waterproofing technologies.

Application Challenge: In-can and Film Microbial Growth

High-build latex emulsion membranes containing hollow glass or ceramic spheres are not structural insulants, but they do slow the rate of heat transfer and make cold masonry surfaces warmer, reducing the risk of condensation and microbial growth. However, water-based/latex emulsions are themselves particularly prone to fungal, algal and bacterial contamination and need additions of a range of biocides with anti-fungal, anti-algal and anti-bacterial properties as preservatives against microbial growth "in-can" during storage, and to prevent microbial growth in subsequently applied "film" coatings.

Microorganisms need both water and nutrients, some, such as algae also need sunlight, and it is the contaminants, bacteria or yeasts in the organic raw materials indigenous to the coatings which act as the microbial food source. If not treated with a broad spectrum of biocides during production, many in-can problems arise, including discoloration, gas generation and foul smells, as well as corrosion of the can/container, rheology changes and, in extreme cases, coagulation of latex emulsions. Dry-film preservatives required to prevent microbial growth of applied coatings may be added during production, or mixed in with the coating immediately prior to application, although this latter approach is not so easy to accomplish with high pigment volume concentration high-build coatings as it is with conventional paints.

As biocidal products are intended to kill living organisms, many present significant risk to human health and welfare, as well as the natural environment, so great care is required when choosing biocides which are regulated in the EU by the Biocide Products Regulations (BPR), and controlled in the USA through listings on the national inventory of chemical substances (TSCA), or are subject to restrictions elsewhere.

Recommended Reading

- *BS EN ISO 1062 Coating Materials and Coating Systems for Exterior Use on Mineral Surfaces and Concrete*, BSI, 2004, ISBN: 0 580 43938 0.

2.3   Roof Protection and Repair

Liquid-based, cold applied, polymeric membrane coatings reinforced with flexible mat, scrim or fleece are used to overlay and protect flat roof plywood boards which have been pre-treated with felt and tar or other waterproofing that has cracked or suffered mechanical damage and sprung leaks; they are equally effective in the repair and renovation of deteriorating profiled sheet metal and asbestos fibre cement pitched roofs. Liquid waterproofing roofing systems also find use in new build projects where various different system approaches are employed depending on lifetime expectations, environmental factors, and roofing design specifics such as resistance to impact damage and foot traffic, as well as complex contours around upstands, skylights, vents or HVAC units. Some technologies are available for emergency leak repairs during adverse weather conditions where there is ponded or standing water, and find use not only on roofs but also on gutters, joints, seams and in open roof valleys. In all situations, cold liquid applied roofing membrane coatings are used because: they are able to follow complex contours; they will fully bond to traditional roofing materials such as felt, asphalt, bitumen, concrete, fiber cement sheet and profiled metal; and because they form a monolithic encapsulating waterproofing membrane and weathering protection. They also avoid the need for hazardous work procedures.

Fact File: Roofing Felt

Bitumen-coated roofing felt is one of the more traditional materials used for covering flat roofs, but it tends to crack on prolonged exposure to UV light or high ambient temperatures. Hot work repair involving torching-on felt patches is an economic but hazardous process due to the significant fire risk to the building, its occupants and contractors; cold work repair involving adhesives to bond patches of felt or other membrane over the cracks is a safer alternative but is not suitable on uneven surfaces or for repairs at cold ambient temperatures. The concept of using natural bitumen liquid as a cold liquid applied waterproofing agent for roofs has progressed significantly over the centuries, and today there are numerous liquid applied roofing technologies which are applied cold, thereby negating any fire risk. When used as overlays on felt, asphalt and bitumen, their lighter colours help reduce heat-ageing stresses.

In addition to their use as overlays to existing waterproofing systems on flat roofs which have cracked, failed at laps and seams, or which have reached the end of their service lives, liquid roofing membranes are also practical repair and maintenance options for balconies and walkways. For reference, new build uses for the same cold liquid-based membrane coating technologies include: warm (also known as built-up) roof systems; inverted (also known as upside down, or buried) roof systems; roof garden (green) systems; and, car park deck waterproofing.

2.3.1 Essential Chemistry and Technology

Cracks, joints and holes in roofing all require sealing or bridging prior to application of a cold liquid-based waterproofing membrane. Sealing of felt, asphalt and bitumen roof defects and joints can be undertaken with a high solids trowel applied mastic based on black bitumen and high levels of reinforcing fibers which dry to a flexible state and remain soft and pliable over the lifetime of the repair.

The alternative approach involves the use of flexible plastic-backed adhesive tape to bridge expansion cracks, holes and joints to provide a smooth surface over which a liquid roofing membrane can be applied. Where bridging tape is used, this is pressed in firmly into the pre-conditioned/primed roofing surface before encapsulation. Conditioner/primer selection is dependent not only on the porosity of the roofing, gutters, concrete/brick upstands, flashings and glazing materials, but also on the selection of liquid roofing membrane.

Historically, liquefied rubbers were used to replace natural bitumen as the basis for the first commercial liquid applied roof coatings since they proved capable of stretching and returning to their original shape without damage, thereby accommodating movement that occurs on many roofs. The traditional natural fibres such as jute and straw, which were added to enhance the strength of the original bitumen liquid coatings, have been replaced by synthetic fibre reinforcements based on glass (random, woven, scrim) or polyester (non-woven bonded or stitched fabric sheet, fleece) to provide a range of weights, strengths and directional strength and elongations needed by the liquid roofing type membrane selection.

Fact File: Synthetic Fibre Membrane Reinforcement

Glass fibres and various types of polyester fibres are used in roofing membrane systems to provide increased strength and durability. Non-woven mats of glass fibres may be made from pulled single filaments or chopped yarns composed of a number of filaments which are then bound or sized together—they are used for enhancing tensile strength, but contribute little or nothing to elongation in an encapsulating membrane. Glass fibres can be woven (intertwined) into mats, or may be formed into a scrim (laid perpendicular and bound)—both types have very good tensile properties (strength rather than elongation) and are used to aid conformability during application of a roofing membrane system. Non-woven polyester mats are made by binding unidirectional extruded polyester filament bundles with thermoset binders, or by loose cross-stitching with themselves—they do nothing for tensile strength, but they do enhance elongation and conformability. Fleeces made from both woven and non-woven polyesters which are needle-punched to create floating surface yarns, have exceptional liquid coating wetting properties and provide omnidirectional reinforcement.

The European (Organisation for) Technical Approvals Guideline (ETAG 005) designates the types of systems currently validated for liquid applied roof waterproofing. The fully endorsed categories include bitumen emulsions and solutions, polymer modified bitumen emulsions and solutions, flexible unsaturated polyester resins, resilient unsaturated polyester resins, single or two-component polyurethanes and water dispersible (acrylic, vinyl-acrylic, styrene-acrylic, styrene-butadiene) polymers and co-polymers. Hot applied polymer modified bitumens are included in ETAG 005, although their use is restricted to inverted/buried roof membrane application or in roof garden/green roof design. Other categories identified, but not covered by the current ETAG, are second-generation styrene block copolymer (SBC) solutions, acrylic polymer solutions, silicones, and methacrylates (sometimes called MMA, PMMA or acrylics).

Application Challenge: Standing Water

Emergency repair of leaking roofs, joints, seams, valleys and gutters requires materials and systems that can be applied and will cure under water. Most liquid roofing systems require all surface water to be removed from surfaces, then cleaning and drying before application with or without any associated primer. In adverse weather conditions it is not usually possible to remove all standing or ponded water from the area of a roof, joint or seam which is leaking from a crack, or to divert running water away from damaged valleys and gutters, so water-tolerant mastic repair materials (also known as flashing cements) are used in what are actually permanent repairs. The mastics are applied by trowel to form a layer over the surface of the crack and a layer of reinforcement sheet embedded into it, followed by another layer of the mastic over the top. High solids, solvent-based single-part rubberised asphalts and bituminous acrylics containing fibres, or high solids two-part plasticised epoxies provide the right level of hydrophobicity required to displace water and permit bonding directly to the underwater surface.

2.3.2 Fit-for-purpose Testing

All roofing repairs need to be watertight and durable under the intended service conditions, and the various roofing systems designated by ETAG have a large number of essential requirements to be met for refurbishment as well as new-build purposes. First and foremost, any repair or refurbishment needs to be resistant to the passage of moisture through the roof into a building so watertightness is evaluated in the EN 1928 test on a free film for resistance to ponding water or to hydraulic pressure absorbed by a limited part of the surface, and also from resistance to water vapour transmission in the EN 1931 test for water vapour permeability.

Roofing systems typically also need to be able to tolerate building movement so, in addition to the standard tests for mechanical durability under tension—ASTM D412 tensile strength and elongation along with ASTM D624 tear strength—resistance to fatigue arising from movement when bridging a gap which opens and closes is verified in the European Organisation for Technical Approval (EOTA) TR008 test where watertightness must be maintained. Assembled roofing systems also need to be able to resist the effects of any likely wind suction/uplift forces and in addition to the conventional ASTM D4541, ISO 4624 pull-off adhesion tests to check for good initial bonding of a membrane and associated primers to various roofing substrates, the resistance to delamination can be verified in the EN 1607 perpendicular-to-face tensile strength test, along with the resistance to wind load by dynamic suction in the EOTA TR005 test method.

The effects of weathering are appraised from changes in ASTM D412 tensile strength and elongation at break properties following ASTM G53, BS EN 1297 ageing/artificial weathering by UV/heat/water exposure with the specific variations suggested in EOTA TR010. Tests which determine the possible effects of accelerated ageing by heat alone, or accelerated ageing in contact with hot water alone, are described in EOTA methods TR011 and TR012, respectively.

The physical durability of a roofing system can be evaluated from resistance to perforation in the EOTA TR006 dynamic indentation test or BS EN 12691, ASTM D2794 rapid impact deformation test, and in the EOTA TR007, BS EN 12730 static load method for watertightness on prolonged contact with a steel indentor. These tests provide an indication of the resistance to foot traffic and loads associated with installation and maintenance of buildings and fixtures such as HVAC units. Also identified as a requirement by ETAG is resistance against plant root growth following the EN 13948 test method.

Probably the most important safety/health/environmental testing requirement for roofing is examination and certification of how it will behave in case of fire. Different countries around the world have their own test method compliance standards with classifications according to external fire performance and reaction to fire. These are governed

by regional, national and international building codes so manufacturers are required to obtain multiple independent certifications
for any given liquid roofing system.

Roofing membranes can be made with solar reflective finishes so
they can be used to reflect the sun's energy and defract any absorbed
infra-red radiation to reduce build-up of heat inside buildings or in the
structures themselves. The EPA Energy Star "Cool Roof" programme
sets performance specifications for solar reflectance as determined in
the ASTM C1549, ASTM E903 or ASTM E1980 test methods.

Recommended Reading

- *Code of Practice – Specification and use of Liquid Waterproofing
  Systems for Roofs, Balconies and Walkways*, Liquid Roofing and
  Waterproofing Association (LWRA), London, UK, 2010.
- *The European Technical Approval Guideline 005 for Liquid Applied
  Roof Waterproofing Kits*, European Organisation for Technical
  Assessment, Kunstlaan 40 Avenue des Arts, B – 1040 Brussels, 2000.
- *ASTM E108-11, Standard Test Methods for Fire Tests of Roof
  Coverings*, ASTM International, West Conshohocken, PA, 2010.
- *BS 476 Fire Tests on Building Materials and Structures. Classifi-
  cation and Method of Test for External Fire Exposure to Roofs*, BSI,
  2004, ISBN: 0 580 43794 9.
- *BS EN 1363, Fire Resistance Tests. General Requirements,* BSI, 2012,
  ISBN: 978 0 580 76353 3.

2.4   Wood Protection

Soft woods, hard woods and engineered wood products are all used
widely as structural materials for the construction of buildings and
boats, as well as for other marine and industrial applications. They
vary in strength, hydrophobicity, density and porosity depending on
the forest timber source—wood is, after all, a naturally occurring
composite with differing cellulose fibre forms and lengths embedded
in varying contents of lignin matrix. Wood properties also change
with heat and/or pressure applied during drying. In the case of wood
products manufactured by bonding together wood particle, fibre,
strand, veneer or lumber laminate with epoxy, phenol-formaldehyde
(PF) or urea-formaldehyde (UF) glues, the resulting engineered wood
components which are generally lighter and stronger than natural
timber, still require waterproofing and protection.

This section describes resinous and polymeric waterproofing and fouling protection for wood in continuous contact with salt water and marine biofouling organisms—the same materials also provide effective barrier protection for wood in fresh, ground and brackish water contact. Waterproofing oils, oil paints and stains that penetrate into the pores and spaces inside wood, or water-repellent preservatives used to combat algae, fungi, lichen, moss, mould or mildew growth on wooden decking are not covered here, nor are wooden cladding paints and stains for buildings requiring protection from rainfall.

2.4.1 Essential Chemistry and Technology

The capacity to form strong bonds between many different kinds of wood, as well as metals and synthetic fibre reinforcements and the ability to penetrate into the grain and pore structures to protect wood against the ingress of moisture resulted in the widespread adoption of two-part liquid epoxy resins in the construction, renovation and preservation of wooden and composite boats. Subsequent to this, liquid epoxy resins have been adopted as the technology of choice wherever primers, undercoats and fillers are required to furnish an effective moisture barrier, as well as a smooth base for other liquid applied polymeric coatings required for aesthetics, foul release or anti-fouling above and below the waterline. At first, two-part epoxy resins required significant dilution with solvents to facilitate application and penetration, but nowadays blends of core bisphenol-A and bisphenol-F diglycidyl ethers with epoxy functional reactive diluents and/or plasticizing non-reactive diluents are used to make liquid epoxy resins fit for saturation of woods and to facilitate formulation as primers, undercoats and fillers with low volatile organic contents.

Fact File: Wood and Glass Reinforced Plastic Boat Hulls

Following the advances made by combining wood with epoxy resin and synthetic fibres in terms of increased capability and resilience, boat building continued to progress and take advantage of the increasing availability of glass, carbon and aramid fibre reinforcement technologies for the development of hulls made exclusively from synthetic fibre reinforced polymer matrix composites. This also involved a change in resin from epoxy to UPR and VE for the polymer matrix to facilitate cost-effective manufacture for speciality performance boat construction. Adversely, GRP (glass reinforced plastic) production boat hulls are prone to absorb water as glass fibre reinforced UPR or VE is porous and the gel coat finishes on the exterior are susceptible to water permeation which can result in **osmotic blistering** over time. Repair of the blistered areas is possible with UPR, VE or epoxy resin based putties, and two-part epoxy barrier over-coatings are invariably applied over them to prevent water migration from the exterior through to the gel coat—these coatings typically contain barrier pigments and fillers of the kind used to protect steel, as described in detail in Chapter 3.3.

After protection with a two-part epoxy moisture barrier/smoothing coating, wooden (and GRP) boat hulls are normally finished with aesthetically pleasing, glossy and UV-stable topcoats made from single-part alkyd, acrylic, acrylic-modified alkyd and silicone-modified alkyd resins, or from two-part PU, acrylic-PU, acrylic-epoxy, and non-isocyanate PU-acrylic resins. For applications where there is limited exposure to sunlight, or where loss of cosmetic appearance over time is not an issue, glossy two-part 'pure' epoxy topcoats can be used. Customarily, one-part topcoats are used where movement of the hull is likely and where regular repainting takes place; two-part topcoats are used to provide longer-term durability, albeit they are less tolerant typically to substrate movement.

Biofouling can be mitigated below the waterline of wooden pleasure craft that are in regular use and removed from the water for frequent cleaning by the application of low surface energy/low coefficient of friction, smooth/non-stick coatings to which marine organisms find difficulty in attaching. Known as foul release coatings, they are typically hard but flexible compositions based on silicone-hydrogel polymers, fluoropolymers, and a variety of proprietary fluorinated coatings including fluorinated-silicone hybrid polymers. None of these technologies require the incorporation of biostats or biocides.

However, optimal protection below the waterline for wooden ships, boats or yachts which are exposed for extended periods to fresh, salt or brackish waters is provided by coatings that are formulated to release biocides and toxins which are lethal to fouling marine organisms. Known as anti-fouling coatings, there are numerous types of polymeric binder approaches and release mechanisms. Historically, tar and oil paints with or without heavy metals or other toxic ingredients were used to protect wooden hulls as well as structures permanently immersed or in tidal zones. Then came biostatic copper protection, and then organotin-based anti-fouling, which, although particularly effective, is now deemed too dangerous for the overall marine environment as the leachates poison non-target organisms such as fish, vegetation, and marine mammals; application is increasingly restricted or prohibited.

Essentially, there are two principal approaches to anti-fouling coating formulation. The first engages hard porous coatings containing active biocides, which leach out in a diffusion-controlled process on contact with water—these are non-erodible and are also known as non-sloughing coatings or hard bottom paints. The second approach involves softer, non-porous coatings which are designed to progressively dissolve in water or erode away to reveal fresh layers of active ingredients locked into the exposed surface of the diminishing protection—they are also referred to as controlled solubility, controlled depletion, ablative, sloughing or self-polishing coatings. Other inventive approaches range from non-stick foul release polymeric binder technology augmented with diffusion-controlled biocidal content to discourage fouling, through to totally biocide-free coatings with synthetic micro-fibers that protrude slightly from the surface and move about in water, preventing growth adhering.

The increasingly inspired combination of binder, particle, fibre and biocide selections to create engineered surface structures which reduce and eliminate the toxicity of anti-fouling, has resulted in the launch of many proprietary polymeric binder technologies in addition to those already used in biocide-containing conventional single-part and two-part topcoat and moisture barrier marine coatings. Where controlled depletion is required, water-soluble resins such as natural gum rosin or acrylic-modified rosins are commonly used: low acrylic/high rosin ratios for soft high eroding coatings; high acrylic/low rosin ratios for harder, slow-release coatings.

2.4.2   Fit-for-purpose Testing

Coatings for wood in continuous contact with water need to bond well, and this can be determined efficiently from ISO 4624/ASTM D4541 pull-off testing of single- or multi-coat primer/build-coat/topcoat systems before and after immersion in water with salinity levels and for periods consistent with expected service conditions and life.

Effective prevention of water migration from the immersed side through to wood relies on minimising the number of voids within a coating system to provide protection from rot or alkaline degradation, which can occur at interfaces with metal fittings, where electrolytic corrosion occurs. Although wooden boats are not susceptible to rusting or osmosis, the fitness-for-purpose of waterproofing barrier multi-coat systems can be verified in the same way as if they were corrosion-resistant coating systems for metal, GRP and even concrete. Straightforward exposure of one-sided coated steel or aluminium panels to water with varying salinity levels following the ASTM D6943/ NACE TM0174 Atlas cell method and evaluation of the degree of degradation as suggested by ISO 4628, along with checks for adhesion loss by, for example, the ASTM D6677 knife test provides a good measure of protection when applied to wood. In addition to salt water contact, marine coatings are also exposed to the effects of sunlight and heat so testing to the ASTM D6695 parameters for sunlight exposure on marine enamels based on the ASTM G151 and ASTM G155 standards permits evaluation of degree of integrity degradation (ISO 4628) and retention of aesthetics in terms of ASTM D523/ISO 2813 gloss and ASTM E1164/BS3900-D10/3 colour retention.

Foul release coatings are typically low surface energy systems designed to minimise the adhesion strength between fouling organisms and a coating surface, and although they do not prevent fouling from settling and growing on a surface, the fouling needs to be easily removed by washing or under hydrodynamic flow, and ASTM D5618, which measures barnacle adhesion strength in shear is used to validate the capability of foul release coating systems to self-clean.

There are a number of methods developed to validate anti-fouling, starting with ASTM D3623, which determines the degree of resistance provided by coatings against attachment of hard (barnacles, oysters) and soft (algae, seaweeds, sponges) organisms in a heavily fouled shallow marine environment. ASTM D5479 is another method used, for both anti-fouling systems and biofouling resistant coatings, involving partial immersion exposure to enhance settlement of fouling organisms and increase the rate of possible physical deterioration.

ASTM D4938 and D4939 are methods specific to anti-fouling on commercial vessels subject to high hydrodynamic stress caused by water flow, and there are a range of test methods which measure release rates of active ingredients: ASTM D5108 for organotin, ASTM D6903 for organic biocides and ASTM D6442 for copper-based anti-fouling.

Recommended Reading

- R. Ciriminna, F. V. Bright and M. Pagliaro, Ecofriendly Antifouling Marine Coatings, *ACS Sustainable Chem. Eng.*, 2015, **3**(4), 559–565.
- C. M. Magina, S. P. Cooper and A. B. Brennan, Non-toxic Antifouling Strategies, *Mater. Today*, 2010, **13**(4), 36–44.
- *Advances in Marine Antifouling Coatings and Technologies*, ed C. Hellio, D. M. Yebra, Woodhead Publishing, 2009, ISBN: 978-1-84569-386-2.

3 Metal Protection and Repair

3.1 Overview of Corrosion and Other Problems That Polymeric Repair and Maintenance Materials Can Solve for Metals and Alloys

Metallurgical science has provided mechanical engineers and materials scientists with an extensive array of metals and alloys which exhibit widely differing and important properties, with steel being the most common structural metal alloy selected for use in the manufacture of equipment, machinery and fittings for industrial production, processing and distribution, as well as for building frameworks and shipbuilding. Steel is the established material of choice where high strength is required and where weight is not a primary concern, as it is long lasting if adequately protected from corrosion—this enduring ferrous metal alloy contains a high proportion of elemental iron, the second most abundant metal in the earth's crust, which is obtained by heating iron III oxide and carbon in the presence of air to extract the base metal in a process known as smelting (eqn (3.1)–(3.4)).

$$2C \text{ (charcoal/coke)} + O_2 \text{ (air)} \gg 2CO \tag{3.1}$$

$$CO + 3Fe_2O_3 \text{ (iron III oxide)} \gg CO_2 + 2(FeO \cdot Fe_2O_3) \text{ (iron II/III oxide)} \tag{3.2}$$

$$CO + FeO \cdot Fe_2O_3 \gg CO_2 + 3FeO \text{ (iron II oxide)} \tag{3.3}$$

$$CO + FeO \gg CO_2 + Fe \tag{3.4}$$

Industrial Polymer Applications: Essential Chemistry and Technology
By William R. Ashcroft
© William R. Ashcroft 2017
Published by the Royal Society of Chemistry, www.rsc.org

Iron alloyed with various proportions of carbon gives low-, mid- or high-carbon steel, and the addition of chromium, nickel and molybdenum to carbon steel produces stainless steels; the addition of silicon, manganese and nickel or chromium to carbon steel results in cast irons and tool steels; so-called alloy steels incorporate these and other elements to meet many other specific design properties. In all cases, the inclusion of carbon and other elements stabilises the movement of dislocations present in the crystal lattice of iron, thereby controlling the ductility, strength and hardness of the base ferrous metal.

3.1.1 General, Uniform, Electrochemical and Wet Corrosion Processes and Mitigation Technologies

All ferrous and non-Noble metals need protection from corrosion. The most prevalent example being **general corrosion** of metals that degrade oxidatively in contact with air and moisture present in the form of condensation, rain or splashing. Also known as **uniform corrosion**, there is always an anodic reaction, and one of two possible cathodic reactions depending on the pH of the exposure environment. Typically, with no preferential site or location for anodic or cathodic reaction, anodes and cathodes form randomly over an exposed metal surface, leading to an essentially uniform loss of metal (eqn (3.5–3.7)).

$$M \gg M^{n+} + ne^- \text{ in the case of iron } 2Fe \gg 2Fe^{2+} + 4e^- \qquad (3.5)$$

$$O_2 + 4e^- + 2H_2O \gg 4OH^- \text{ in alkaline or neutral environment}$$
$$(\text{pH} = 7 \text{ or pH} > 7) \qquad (3.6)$$

$$2H^+ + 2e^- \gg H_2 \text{ (gas) in an acidic environment (pH} < 7) \qquad (3.7)$$

Given sufficient time and oxygen/water exposure, ferrous metal exposed to the atmosphere is known to revert to its lowest energy state, converting entirely to rust and disintegrating through the uniform corrosion process. Specifically, iron as anode dissolves from under a wetted surface, forming ferrous hydroxide ($Fe^{2+} + 2OH^- \gg Fe(OH)_2$), leaving a rough surface with a loosely bound porous coating of precipitated ferrous hydroxide (known as green rust), which oxidises to ferric oxide–hydroxide ($FeO(OH)$) and ferric hydroxide ($Fe(OH)_3$), which ultimately transforms to hydrated ferric oxide (red rust, $Fe_2O_3 \cdot nH_2O$). During dry periods, when the oxidation products dehydrate completely, the resulting red powdery ferric oxide can be lost to wind erosion, exposing bare metal with an increased surface area with even higher susceptibility to the corrosion and oxidation process when moisture returns.

In submerged situations where drying out does not occur, corrosion products develop on the surface of ferrous metals and form crystalline tubercules that roughen the metal surface and are themselves susceptible to erosion in situations where there is water flow. In the case of seawater immersion or exposure, more serious **electrochemical corrosion** or **wet corrosion** results due to saltwater being a highly effective electrolyte for the promotion of electron transfer from the anodic area within the electrochemical corrosion cell to the cathodic area where dissolved oxygen gets reduced to hydroxide ion (Figure 3.1).

Natural mill scale, which is the thin but tightly bound layer of iron oxides that form on the outer surfaces of rolled steel, provides natural protection from atmospheric/uniform corrosion for as long as there are no breaks in the scale. This works because corrosion is a diffusion-controlled process, and any method that limits access of oxygen and moisture, or reduces the activity of an exposed surface, will improve a material's corrosion resistance.

Although the activity of non-Noble metal surfaces can be reduced by passivation, with ferrous metals it is usual to remove mill scale and keep from rusting by one or more methods of galvanic protection. Galvanising or protection with a plating of zinc, which acts as a sacrificial coating is useful where there will be no exposure to

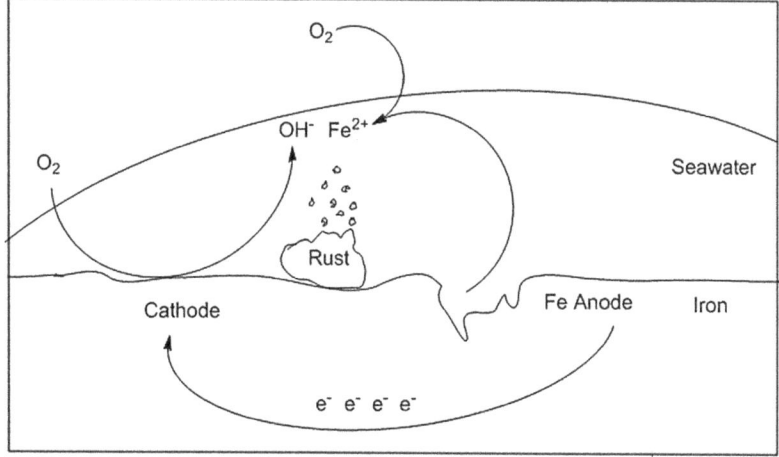

Figure 3.1 Electrochemical corrosion cell for metal exposed to seawater.[1]

corrosive materials such as salt water. Alternately, high-build zinc-rich primers are used to protect iron and steel as the zinc in a primer will corrode sacrificially in preference to the underlying ferrous metal, even when an inert coating or paint applied over them gets damaged by scratching or abrasion—they are also useful for repairing damage to galvanized steel. Cathodic protection is a technique that makes the metallic structure the cathode of an electrochemical cell by application of an impressed current or by connection to a sacrificial galvanic anode—in both cases, the metal also requires a protective coating system to provide a barrier against oxygen and water and to provide other functionality, depending on intended purpose and environmental exposure.

Other metals undergo equivalent corrosion processes creating oxides at the surface—furthermore, exposure to carbon dioxide, acid rain or other sulfurous atmospheric elements leads to the formation of carbonates, sulfates and sulfides in addition to oxides, which results generally in the formation of a patina or tarnish on a metal's surface. Metals deliberately exposed to etching acids undergo what is effectively another type of uniform corrosion process where corrosive attack proceeds evenly over the surface of exposed metal. Most types of corrosion are, however, unintentional and highly localised, with mechanisms and processes that are less predictable and less obvious. However, before other forms of corrosion and preventive mainten-ance requirements are considered, the characteristics of some other important non-ferrous metals and alloys that also get specified for equipment, machinery, fittings and structural applications need to be considered.

Non-ferrous metals and alloys do not contain iron and are therefore non-magnetic. Although pure non-ferrous metals are rarely used as structural materials as they lack mechanical strength, they are widely used in alloys with other metals and elements that improve their strength. Aluminium, which is even more abundant than iron on our planet, is widely used because it is light, pliable and rust-resistant.

Aluminium metal is extracted from its ore (alumina Al_2O_3) by electrolysis as there is no commercially practical reducing agent to smelt it—hence its resistance to atmospheric oxidation. Aluminium is alloyed typically for structural applications requiring the highest strength-to-weight ratios and toughness at extreme temperatures, with alloying elements including copper, magnesium, manganese, silicon, tin and zinc, all of which also enhance resistance to chemical and oxidative corrosion under forcing conditions.

Copper alloys have been used by engineers for a long time for their resistance against corrosion, and brass (made from copper and zinc) modified with added arsenic or tin makes corrosion-resistant brasses for equipment and fittings used in harsh environments and where low friction is critical; alloys made from copper, zinc and aluminium (known as aluminium brass) are used for heat exchanger and condenser tubes. Bronze (made from copper and tin) modified with added phosphorus (known as phosphor bronze) is used to cast ships' propellers and make low friction bearings. Aluminium bronze (made from copper and aluminium) is very hard and used for bearings and machine tooling, as well as propellers and underwater fastenings on ships. The presence of copper in alloys also introduces important biostatic effects—it stops bacteria from reproducing while not necessarily harming them, and resists unwanted colonisation by marine organisms including barnacles, mussels and algae—making them preferable to steel and other non-cupric metals in marine environments.

3.1.2 Other Insidious Corrosion Processes and Mitigation Technologies

In addition to the highly visual problems caused by the uniform corrosion process, there are numerous other dangerous corrosion processes that will degrade and destroy metals and alloys if left unprotected.

Bi-metallic corrosion—non-ferrous metal alloys clearly have their uses as specialist engineering materials and are widely specified where iron and steel have to be avoided. Paradoxically, non-ferrous metal components are often used in conjunction with structural iron and steel materials, and where they are coupled or jointed can lead to bimetallic corrosion, also known as contact or galvanic corrosion. Where two different metals are in electrically conducting contact, and their individual electrochemical potentials are sufficiently far apart, the baser metal will corrode more quickly than if it were isolated from the nobler one—this is, of course, the premise of cathodic protection where the aim is to protect the nobler metal from oxidation—and coupling components containing nobler metal elements than iron will promote rusting of the structural steel. The risk of corrosion increases in the presence of an electrolyte (corrosion salts which dissolve/ionise in water, or seawater) but can be mitigated with a protective coating to provide a continuous, inert and adherent film over the contacting metal surfaces and between the environment by preventing liquid electrolyte permeation.

In terms of less predictable or less visible corrosion processes which may attack metals and their alloys, ductile metals with residual stresses from manufacture or that are placed under static tensile stress in a corrosive environment at elevated temperatures can suffer significant loss of mechanical strength and catastrophic failure in a form of corrosion called **stress corrosion cracking** (SCC). This type of corrosion is not obvious from casual inspection, as there is little associated metal loss and so it can go undetected prior to failure. There are a number of causes of SCC initiation (active path dissolution, hydrogen embrittlement and brittle film-induced cleavage) and prevention requires eliminating one of three contributing factors. This can be achieved by: changing to a material that is not susceptible to SCC in the service environment; reduction of stress in the material below the threshold stress for SCC; or more simply isolating the metal from the corrosive environment—when this is done with a surface coating, care in selection is required as film-induced cleavage of the metal can arise from cracks initiating in a coating if it in turns becomes brittle in the corrosive environment.

Another deceptive corrosion process that may go undetected prior to malfunction is known as **selective dissolution corrosion**—this can occur in two-phase alloys and involves the baser metal phase dissolving and leaching out selectively, leaving the nobler metal phase intact but with significant porosity and reduced mechanical integrity. The risk of selective dissolution corrosion is typically mitigated by heat pre-treatment to fortify the crystallite microstructure, and either or both cathodic protection and isolation with a protective coating.

Intergranular corrosion is a form of localised corrosion in metal alloys with susceptible grain boundary regions that develop into galvanic couples, leading to local galvanic corrosion in the presence of electrolytes. The problem occurs when the alloy sees exposure for long periods at temperatures of 500–800 °C (900–1500 °F) and cannot be eliminated by a coating. When intergranular corrosion arises due to welding the corrosion is termed weld decay which can be controlled by grinding to remove corrosion deposits from the weld and protection with a polymeric surface coating which covers both the weld and parent metal to isolate them from the environment This only works, however, if the equipment service temperature does not exceed the capability of the protective coating material.

Localised corrosion is characterised by accelerated attack at small areas or zones on a metal surface in contact with a corrosive environment, and can easily be confused with uniform corrosion when it manifests itself over large areas. As already discussed, bimetallic, intergranular and weld decay are all forms of localised corrosion, and there are a number of other common localised corrosion mechanisms affecting metals—for example, pitting, crevice and filiform corrosion—which all lead to damage that fortunately can be repaired with polymeric materials.

Pitting corrosion is a highly localised form of attack occurring in the bottom of microscopic surface anomalies or pits in a metal. Corrosion involves destruction of the passive layer in the bottom of the pits, creating small corrosion anodes connected with surrounding bulk metal, which becomes a very large cathode in a galvanic corrosion cell. Hydrolysis of the corrosion products that form in the bottom of the pits leads to a drop in pH, which increases the polarisation and the rate of corrosion forming deeper and deeper pits, whilst most of the surrounding metal surface shows no corrosion at all.

Pitting corrosion is made worse in the presence of chlorides and by rising temperatures, and can result in perforation of the walls of pipes, tanks, valves and pump casings in a relatively short time period.

All common metals and alloys, including stainless steel and aluminium alloys, are susceptible to pitting corrosion depending upon the conditions of service, but especially during equipment outages when left in contact with stagnant solutions with depleted oxygen levels—the corrosion resistance of stainless steel depends upon continual access to oxygen on a wetted surface to maintain the protective oxide film. The problem can be addressed in advance by using steels alloyed with molybdenum, with the amount required to ensure protection being increased according to the expected chloride contact and temperature; or by repairs with cold curing polymeric repair composites (also known as plastic metals), polymeric coatings and cathodic protection as appropriate when hot-work welding is not a viable option.

Crevice corrosion is essentially another form of pitting corrosion, and a further localised type of attack occurring in gaps or fissures underneath surface encrustations and weld deposits on metal surfaces, or between two metals or a metal and non-metallic material (such as a gasket) in close contact with one another—anywhere in practice that a thin film of water can penetrate and the availability of oxygen is low. Most metals and alloys are susceptible to corrosion inside crevices which fill with liquid that will stagnate, and the severity of corrosion increases with the narrowness and depth of the crevice; this problem is not found in grooves or slots in which circulation of liquid is possible. This form of corrosion is a particular problem with stainless steels, especially when in contact with liquid containing chloride salts, which can concentrate inside a crevice and gradually acidify the stagnant solution, leading to highly aggressive local conditions which destroy the passivity—any metal with corrosion resistance based on its capacity to form a protective oxide (passive) film is susceptible to localised corrosion when the passive film breaks down. As with pitting corrosion, there are highly resistant alloys which can be used to repair and replace components susceptible to crevice corrosion, and there are additional corrosion control measures involving cathodic protection, sealing with an effective polymeric barrier coating, or the use of corrosion inhibitors in the liquid medium contacting uncoated metal. Other planning changes include elimination of "dead legs", where liquid can stagnate, ensuring full drainage during prolonged out-of-service periods, improving joint and gasket design, and also allowing for easier cleaning of surface deposits that collect during service.

Filiform corrosion, also known as underfilm or wormtrack corrosion, is a type of crevice corrosion that occurs on metallic surfaces under painted or plated surfaces in contact with seawater. Corrosion starts at pinholes or defects in a coating from scratches or impact damage, with lacquers and quick-dry paints being particularly susceptible. Where the metal is exposed, soluble metal chloride salts form and the oxide corrosion products grow in the shape of filaments anodically undermining the coating; oxygen and water vapour diffuse through the so-called filiform tail and drive the cathodic reaction of the metal just below the coating at the front of the filament in a self-propagating process. Growth is more vigorous under thicker organic films than thinner films, and in practise filiform corrosion is minimised by careful surface preparation to eliminate defects prior to coating, or by the use of coatings that are resistant to corrosion, and by careful inspection to ensure pinholes or holidays in the coating are covered by additional coats where necessary.

Microbiological influenced corrosion (MIC)—microbial or bacterial corrosion effects metals as well as non-metallic materials like concrete as previously mentioned in Section 1.7. It is caused by bacteria in biofilms feeding off nutrients and oxygen (although some types of bacteria need only very small amounts of oxygen) on the corrodable inside surface of tanks, pipes, equipment or systems where water is enclosed. There are various types of microbes, which tend to act synergistically, leading to corrosion by production of corrosive species such as mineral acids and organic acids, ammonia, sulfides (by reduction of sulfate), or through the formation of differential aeration corrosion cells. Control of MIC is achieved by water treatment, addition of biocides and corrosion inhibitors, and by protection of the metal with a MIC resistant protective polymeric barrier coating.

3.1.3 Problems Other Than Corrosion That Polymeric Repair and Maintenance Materials Can Solve

In addition to controlling the ravages of corrosion, there are a number of important other ways that polymeric materials can be used to protect and repair metals: as sacrificial metal replacement for eroded (and corroded) areas; the repair of cracks and holes; the creation of irregular shims; and for bonding metal components. The choice of polymeric coating or repair composite depends on how the problems arise and the likelihood of recurrence, the options and details for which are covered in the following sections.

Recommended Reading

- *Mechanical Engineer's Handbook*, ed. M. Kutz, John Wiley & Sons Inc., New York, 3rd edn, 2005.
- *Corrosion Handbook,* John Wiley & Sons Inc., 1999–2014, Online ISBN: 9783527610433.

3.2 Corrosion-resistant Primers

Zinc-based corrosion protection is the most versatile and effective means of prevention or delay of rusting of iron and steel, as it can be applied at point of manufacture or post-manufacture after transit and assembly on-site. It can be applied in the form of: an elemental zinc coating by hot-dip galvanising or by thermal-/flame-spraying; as a zinc phosphate conversion coating; or alternatively in the form of high-build zinc-rich paints and primers. This section covers the use of zinc amongst other reactive anti-corrosive fillers and pigments when formulated into liquid applied coatings and primers for steel—Section 3.3 deals with barrier pigment filled corrosion resistant polymeric coatings required to seal them as well as for sealing thermal spray zinc, aluminium and their alloys used to protect structural steel from general and localised corrosion.

3.2.1 Essential Chemistry and Technology

Ultra-fine zinc dust incorporated into a binder will act as a liquid anodic coating, providing there is a sufficiently high percentage of zinc in the dry film after cure for the particles to remain in contact with each other. The binders can be either inorganic or organic in nature, with levels of zinc in the dry film typically in excess of 90 wt.% for zinc-rich paints which are used for repairing damaged galvanised coatings and also in their own right for uncoated steel as they possess similar characteristics to a hot-dip galvanized zinc coating. Zinc primers with zinc dust contents in the dry film of 25–70 wt.% are used to protect steel in conjunction with barrier coatings which decrease permeability to oxygen and water, and where necessary also with a finish coat over the intermediate coat to provide the required appearance and surface resistance of the system against weathering or other exposure conditions.

Zinc primers made with inorganic silicate binders are known as zinc silicates and are either made from alkali metal (sodium, potassium or lithium) silicates which are waterborne, or from ethyl silicates which are solvent-containing—they "moisture-cure" to a porous glass silicate matrix which locks metallic zinc in position. Zinc ethyl silicates are the most common selection due to their wider compatibility with topcoats, their extended durability and also as the preferred choice for shop primers which are thinly applied at 10–15 μm (0.4–0.6 mil) for the temporary protection of steel sheets—they contain no organic materials or alkali metals when cured and do not interfere with welding and cutting processes. Organic zinc primers made with waterborne crosslinking epoxy resins can be either two-part or three-part systems (with zinc dust as one of the three parts), or are made from solvent containing crosslinking epoxy, air drying epoxy ester, or moisture cured PU resins. The waterborne alkali metal silicates are the more environmentally friendly option as they can be made with zero VOC unlike waterborne epoxies, which need co-solvents to ensure good film formation. Figure 3.2 presents a comparison of the environmental impact of the various solutions from an installation perspective.

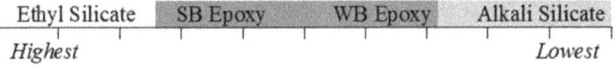

Figure 3.2 Environmental impact during installation of corrosion-resistant primers.

Waterborne zinc-rich and reduced zinc primers typically contain anti-gassing inhibitors to prevent the emission of hydrogen by protecting zinc from reacting with water during dispersion. These and the solvent containing types include anti-settling and thixotropic agents to prevent or minimise separation of the zinc dust in-can.

The alternative to zinc-rich or zinc-reduced primers that are anodic to steel, are primers based on binders that bond well to the metal substrate and which include inhibitive and non-inhibitive barrier pigments to interfere with the corrosion process. Highly effective inhibitive pigments such as white/red lead and metal chromates have been replaced with safer alternatives, including red iron oxide, zinc phosphate, aluminium tripolyphosphate, and ion-exchanged synthetic amorphous silica technologies. Micaceous iron oxide (MIO) which is dark grey in appearance with a metallic sheen differs in form, shape and function from red iron oxide—MIO is flaky and lamellar in structure (called "micaceous" because the lamellar particles are similar to mica) and is widely used as it acts an effective type of anticorrosive barrier pigment for reducing or delaying moisture penetration in high humidity environments.

Fact File: Zinc Phosphate

Zinc phosphate is a very versatile inhibitive pigment and commonly used in primers that are applied directly onto steel, or in the case of duplex systems onto a sealed metal coating. Zinc, along with manganese and iron phosphates, are also used for phosphating steel in a process involving immersion or spraying with a dilute solution in phosphoric acid to create a conversion coating of insoluble, crystalline phosphate formed by chemical reaction with the surface—this creates lubricity as well as corrosion resistance, although conversion coatings are porous and need sealing or coating to prevent moisture and oxygen ingress.

Single-part air drying alkyd resins are the binder of choice where protection of internal or external steel structures in light industrial environments is required; rubber modified alkyds and single-part film forming vinyls, chlorinated/acrylated rubbers are used for immersion duty; two-part epoxy and PU resins provide the highest strength adhesion to metal as well as strong cohesion and bonding to pigments forming additional passive barrier protection for heavy/high-duty protection. Liquid epoxy resins of low viscosity allow the formulation of solvent-less and solvent-free systems for application as very thick films, although they are the most difficult to apply and

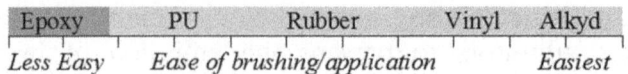

Figure 3.3   Ease of brushing/application of corrosion-resistant primers.

require heated two-component spraying equipment. Figure 3.3 provides a qualitative assessment of the ease of use of single-part and two-part corrosion resistant primer technologies.

Single-component moisture-cured PUs containing non-leafing aluminium find use as corrosion barrier primers and as tie-coats for many coating types—they are favoured for their exceptional adhesion to sound, tightly adherent rusty steel, and other marginally prepared surfaces.

3.2.2   Fit-for-purpose Testing

Primers must adhere to the substrates on which they are applied, and also to coatings applied over them when used as part of a multi-coat system, and there are two published standards of relevance in the verification of fitness-for-purpose of corrosion-resistant primers: ASTM D3322, which covers testing of primers and primer surfaces over pre-formed metal, and ASTM D1654, which deals with testing of coating systems subjected to various simulated corrosive environments.

Test methods recommended for adhesion are the ASTM D3359 tape test, the ASTM D2197 stylus scrape test, and the ASTM D6677 knife test which, even though not providing any quantitative data, is the more informative method for establishing the adhesion of a primer to metal and to another coating in a multi-coat system.

There are a number of traditional laboratory methods for simulating different corrosion environment exposures. ASTM B117 and ISO 11997 involve not only continuous spray of an atomised salt solution (salt fog) to represent conditions encountered in a pure marine environment, but also cyclic corrosion conditions of wet (salt fog)/dry/humidity/UV light as often occurs in natural conditions.

ASTM G85 Annexe A5 involves wet-dry cycling, with a spray of atomised salt and ammonium sulfate solution in what is known as the prohesion test to represent exposure to industrial, as well as marine and coastal conditions, where acid rain contributes additional challenges. ASTM D4585 and ISO 6270 deal with constant condensation-water atmospheres and alternating condensation-water atmospheres to represent humid conditions and condensation arising by settling of fog and mist and temperature fluctuations below the dew point. Test panels are normally scribed with an "X" as per the ASTM D1654 standard for evaluation of coating panels subjected to corrosive environments, with comparative evaluations of performance made by examination of the degree of corrosion staining, blistering associated with corrosion, loss of adhesion at the scribe mark or other film failure. None of these tests can be used to determine an actual service life, and need to be run for extended periods before even comparative evaluations of performance can be made.

In contrast, the ISO 17463 standard involves an accelerated cyclic electrochemical technique (ACET), which allows comparison of the protective and anticorrosive properties of coating systems on metal within short time periods in a qualitative and quantitative way. It includes cycles of EIS (electrochemical impedance spectroscopy) measurement, cathodic polarisation and potential relaxation, which provide an evaluation of the permeability of the coating and properties which can be attributed to adhesion to the substrate. Figure 3.4 provides a qualitative assessment of the resistance to corrosion of single-part and two-part corrosion-resistant primer technologies.

ASTM D2803 is the guide for testing filiform corrosion resistance of coatings.

Application Challenge – Cathodic Disbondment

Cathodic protection systems are used to prevent corrosion of structural metals by reducing the electrical potential below their corrosion potential. However, impressed currents applied to a metal can lead to disbondment of a coating system where corrosion occurs underneath the coating. Current will pass through breaks or defects in a coating, and if the size of the fault increases then more current will flow through, resulting in a voltage drop at the interface with the coating being forced away from the metal as hydrogen gas bubbles develop and ferrous hydroxide precipitating at the metal surface. The ISO 15711 and ASTM G8/G42 methods are used to determine the compatibility of coating systems with cathodic protection (ISO 15711 for coatings exposed to sea water; ASTM G8/G42 for pipeline coatings). Other factors affecting disbondment of applied coating systems include thickness as well as uniformity/freedom from flaws, and also degree of cure/ageing.

Alkyd, Vinyl, Rubber	Epoxy, PU	Zinc Epoxy	Zinc Silicates
Good			*Excellent*

Figure 3.4 Corrosion resistance of single-part and two-part primer technologies.

Recommended Reading

- *BS5493 Code of Practice for Protective Coating of Iron and Steel Structures Against Corrosion,* BSI, 1977, ISBN: 0 580 09565 7.
- R. Hudson, *Coating for the Protection of Structural Steelwork,* Department of Trade & Industry/National Physical Laboratory UK, 2000, download from www.npl.co.uk.
- *ISO 12944 Corrosion Protection of Steel Structures by Protective Paint Systems – Part 5: Protective Paint Systems,* International Organization for Standardization, Switzerland, 2007.
- *ISO 20340 Performance Requirements for Protective Paint Systems for Offshore and Related Structures,* International Organization for Standardization, Switzerland, 2009.
- M. O'Donoghue, V. J. Datta, M. Winter and C. Reed, *J. Prot. Coat. Linings,* 2010, 30–45.

3.3   Corrosion-resistant Barrier Coatings

This section deals with polymeric coatings containing barrier pigments and fillers, which are used alone or in combination with primers and sometimes finish coats, for the protection of structural steel, concrete and masonry from general corrosion. Coatings for the protection against erosion/abrasive attack are dealt with in Section 3.4; coatings for the protection against aggressive chemical attack are discussed in Section 3.5.

3.3.1   Essential Chemistry and Technology

Conventional corrosion-resistant coating systems for low-alloy carbon steel comprise a primer, undercoat and a finish coating applied to metal which is degreased and blast cleaned to SSPC-SP10/NACE No.2/ISO 8501/BS 7079 Sa $2\frac{1}{2}$ standards minimum, or which has been mechanically prepared by chipping, scraping and wire brushing to remove loose millscale and rust to SSPC-SP3/ISO 8501/BS 7079 St 3 standards minimum.

Fact File: Surface Preparation Standards

There are numerous national and international standards relevant to the preparation of the surface of steel substrates before the application of industrial coatings and related products:

- **SSPC-SP1** Solvent cleaning—requires the removal of oil, grease, dirt, soil, salts and contaminants;

- **SSPC-SP2, ISO 8501/BS 7079 St 2 or St3** Hand tool cleaning—requires the removal of loose coating, loose rust and loose millscale by hand chipping, scraping, sanding and wire brushing;
- **SSPC-SP3, ISO 8501/BS 7079 St 2 or St3** Power tool cleaning—requires the removal of loose coating, loose rust and loose millscale by power tool chipping, descaling, sanding, wire brushing and grinding;
- **SSPC-SP5, NACE No.1, ISO 8501/BS 7079 Sa 3** White metal blast cleaning—requires the removal of oil, grease, dust, dirt, coating, mill scale, rust or other corrosion products, and other foreign matter by blast cleaning prior to the application of protective coating systems for the severest corrosive atmospheres and immersion service;
- **SSPC-SP7, NACE No.4, ISO 8501/BS 7079 Sa 1** Brush-off blast cleaning—requires the removal of oil, grease, dirt, dust, loose coatings, loose mill scale and loose rust, leaving only tightly adhering residues of coatings, millscale and rust;
- **SSPC-SP10, NACE No.2, ISO 8501/BS 7079 Sa $2\frac{1}{2}$** Near-white blast cleaning—requires the removal of oil, grease, dust, dirt, coating, millscale, rust or other corrosion products, and other foreign matter from at least 95% of the surface area prior to the application of protective coating systems for corrosive atmospheres and immersion service;
- **SSPC-SP14, NACE No.8** Industrial blast cleaning – requires the removal of oil, grease, dust, dirt, coating, millscale, rust or other corrosion products, and other foreign matter from at least 90% of the surface area prior to the application of protective coating systems.
- **SSPC-SP WJ-1,2,3,4, NACE WJ-1,2,3,4** A choice of more environmentally friendly cleaning methods compared to conventional dusty dry abrasive blasting methods—ranging from low-pressure or high-pressure water cleaning to high-pressure or ultra-high-pressure waterjetting, which are used to prepare surfaces for recoating by: light cleaning to permit tightly adherent coating, millscale, rust or other corrosion products to remain (WJ-4); a more thorough cleaning (WJ-3); a very thorough cleaning (WJ-2); or for the complete removal of all coatings, millscale, rust or other corrosion products to bare substrate and exposure of an original abrasive-blasted surface profile (WJ-1).

Primers, as already discussed in Section 3.2, contain reactive/ sacrificial fillers and pigments to scavenge electrons and chloride ions; undercoats (or intermediate coats) are used to 'build' barrier film thickness and are often applied in several coats; finish coats (or top coats) provide the required appearance and surface resistance of the system. Primer/undercoat/finish coats may all be the same generic type of coating material, or they may all be different, and are characteristically solvent containing to facilitate speed and ease of application by brush, roller or spray, as well as physical drying—ISO

12944, which is the global benchmark for protection of low alloy carbon steel from corrosion in the atmosphere/water/soil by protective coating systems, defines fit-for-purpose combinations for typical exterior and interior corrosive environments. Typical options for high build undercoats are zinc phosphate alkyd, zinc phosphate/MIO/glass flake epoxy, or glass flake UPR and VE.

Fact File: Finish-/Top-coat Technologies

Finish-/Top-coat resin technologies for corrosion resistance include:

- Air-drying alkyds, urethane alkyds and epoxy ester enamels which provide general purpose coatings with good overall corrosion resistance;
- Single-part physical drying acrylics, acrylated rubbers, chlorinated rubbers and vinyl chloride copolymer (vinyl) resins which provide a higher level of resistance to atmospheric pollution and immersion in salt and fresh water;
- Two-part PUs which provide gloss and colour retention as well as chemical and corrosion resistance;
- Two-part epoxies based on low viscosity liquid resins that do not need a solvent to give outstanding corrosion resistance in the most aggressive chemical environments.

All of these coating types are also available typically with added solvented asphaltic bitumen or coal tar pitch for cost reduction and corrosion resistance enhancement.

Corrosion-resistant coating systems for concrete and metal, which can be used without a corrosion-resistant primer are either: single thick coats of elastomeric PU or polyurea; one to two coats of high-build glass flake epoxy/UPR/VE; or two to three coats, each pigmented a different colour to ensure no misses and to safeguard against coating holidays, of epoxy or PU containing liquid resinous extenders in addition to barrier pigments and fillers.

All these high-build coating types are characteristically solvent-free and more demanding to apply by brush or airless spray and are intended for use on metal which has been degreased, blast cleaned to ISO 8501 standard Sa $2\frac{1}{2}$/SSPC-SP10 minimum (rather than manually prepared to ISO 8501 St 3/SSPC-SP3 minimum), and surface salt contamination carefully removed. In contrast, concrete and masonry requires light abrasive blasting to remove laitance, dirt or previous coatings with holes and minor irregularities in the substrate being filled with an epoxy fairing or smoothing coat to provide an impermeable seal before coating.

Application Challenge – Salt Contaminated Surfaces

There are a number of sources of salt contamination that are dangerous to low alloy steel on exposure in marine and industrial environments: chlorides are the most common on structural components which have been immersed for any period in seawater or left out in the open in a marine or chemical treatment industrial location; sulfates and nitrates can also deposit from the atmosphere on the surface as well as in ground water in environments where there is heavy industrial activity and high automotive traffic density. When present in sufficient concentrations, soluble salt contamination leads to premature coating failure, particularly in humid or immersed hot immersion applications. Salt contamination and brine deposits, which are particularly hygroscopic, trapped underneath a coating attract water by osmosis. Water permeating through a coating, especially through any porosity, builds up at the interface between steel and a coating, leading to the creation of osmotic cells, which pull in more moisture to dilute the concentrated solution creating blisters under the coating which then typically disbonds—electrolytic cells also form where oxygen permeates leading to under-film corrosion and further blistering/delamination.

Blast cleaning of salt-contaminated surfaces before coating application increases the problem by forcing salts into the metal surface unless washing the surface with deionised/distilled water is included in the cleaning process. After degreasing and blasting to the required surface profile, metals need to be allowed to stand to allow any ingrained salts to sweat to the surface and then they should be washed prior to a further brush blast to remove them. With dry blasting, the process may need to be repeated several times to ensure sufficient removal of salt contamination from the surface. Wet abrasive blasting or ultra-high-pressure water jetting can remove high proportions of soluble salts but in locations where there is a shortage of clean deioinised water, it may be necessary to include the use of a proprietary salt-removal treatment in the cleaning process prior to a final distilled water rinse and final blast.

Tests for the assessment of salt contamination on blast-cleaned metal surfaces are covered in ISO 8502-2, which describes methods that identify total salt levels, specific chloride levels, and levels for less aggressive sulfates and nitrates. ISO 15235 provides guidance on permissible levels of water-soluble salt contamination for seawater immersion and marine exposure applications, and in general the maximum amount of salt allowed on a surface prior to coating is determined typically by the coating supplier as acceptable thresholds for salt contamination decrease significantly as service temperatures increase.

Both conventional intermediate coats for steel and high-build coatings for concrete, wood and metal incorporate barrier pigments and fillers to decrease permeability to oxygen and water. They include: lamellar or plate-like pigments and fillers such as MIO, glass flake and talc; needle-like crystalline silicate minerals such as calcium metasilicate and fibres such as ceramic, glass or synthetics; and spherical fillers such as glass beads or microspheres. They are typically used in conjunction with each other and with sacrificial pigments to maximise packing and increase tortuosity of the path length through the pore system in the applied coating system. Liquid resinous extenders are used to improve resistance to water and salt solutions, and as they do not react with the binder components their addition levels are limited to avoid adverse effects on reactivity (extension of drying/curing rates) and film properties (lowering of hardness/over plasticisation)—non-toxic synthetic hydrocarbon resins are now used typically in preference to coal tars formerly used extensively.

3.3.2 Fit-for-purpose Testing

Testing of corrosion-resistant coating systems based on high-build barrier coatings, with or without a corrosion-resistant primer or a finish coat, requires verification of durability on exposure in a wide variety of simulated atmospheric and immersed corrosive environments depending upon intended service. Protective coatings for offshore structures in particular require testing not only for the zone submerged below sea level, but also the splash zone and the atmospheric zone not in contact with seawater.

The traditional atmospheric corrosion simulation tests described for corrosion resistant primers in the previous section also apply and are: ASTM B117, ASTM D5894, ISO 9227, ISO 11997 continuous salt fog spray and cyclic salt fog/dry/humidity/UV light exposure; ASTM G85 A5 prohesion; and ASTM D4585, ISO 6270 continuous condensation. In addition, ISO 20340 calls for cyclic salt spray/QUVB/sub-zero temperature exposure to evaluate the specific suitability of protective coatings applied to offshore and related structures.

For corrosion-resistant barrier coating systems used as immersion linings or submerged in service, ASTM D6943/NACE TM0174 standards encompass useful test methods involving completely encapsulated panels and also one-sided panel testing in Atlas cells involving contact with chemicals in the liquid and vapour phase—one-sided panel testing permitting evaluation of the permeability and cold wall blistering potential of a lining in addition to chemical resistance (see also Section 3.5). Immersion corrosion test panels are not normally scribed as is required for non-immersion coating systems, and comparative evaluations of performance advocated by ISO 4628, include examination of the degree of blistering (ASTM D714), rusting (ASTM D610) as well as checks for cracking, flaking and film lifting, softening or destruction. Checks for loss of adhesion are made by the ASTM D3359 tape test, the ASTM D2197 stylus scrape test or the ASTM D6677 knife test. In the case of immersion coatings tested in an Atlas cell, pull-off adhesion testing in accord with ASTM D4541 and ISO 4624 is used to verify performance in both the immersed and vapour phase to check for susceptibility to liquid and vapour absorption, leaching or other degradative attack. Figure 3.5 provides a qualitative assessment of the resistance to corrosion of single-part and two-part barrier coating polymer technologies.

The ISO 17463 accelerated cyclic electrochemical test standard is again highly useful for comparative evaluations of protective and anticorrosive properties, and coating impedance measurements determined from static electrochemical ionisation spectroscopy (EIS) testing as prescribed in ISO 16773 provide good correlations with performance in service. Figure 3.6 presents a quantitative summary of corrosion protection based on coating impedance measurement.

Figure 3.5 Corrosion resistance of single-part and two-part barrier coating technologies *especially when modified with coal tar and tar replacements.

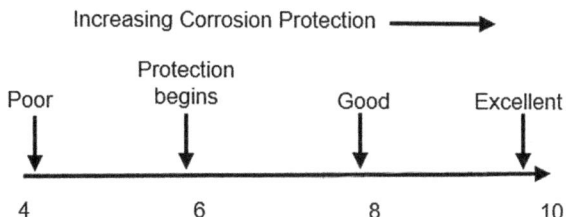

Figure 3.6 Corrosion protection as a measure of coating impedance Log Z (Z in ohm cm^2 @ 0.1 Hz).[2]
Reproduced with permission from NACE International, Houston, TX. All rights reserved. L. G. S. Gray, B. Drader, M. O'Donoghue, R. Garrett, R. Graham and V. J. Dratta, Using EIS to Better Understand Tank Lining Performance in Laboratory and Field Evaluation, Corrosion 2003, paper 03382, 2003. © NACE International 2003.

Recommended Reading

- *ISO 8501 Preparation of Steel Substrates Before Application of Paints and Related Products,* International Organization for Standardization, Switzerland, 2007.
- *ISO 12944 Parts 1 to 8 'Paints and Varnishes – Corrosion Protection of Steel Structures by Protective Paint Systems',* International Organization for Standardization, Switzerland, 1998 to 2007.
- *M-501 Surface Preparation and Protective Coating,* Norsk Standard, Norway, 6th edn, 2012.

3.4   Erosion-resistant Coatings

Metals and alloys, just like concrete, masonry and other engineering materials, are vulnerable to surface erosion from mechanical impact and the cutting action or abrasive wear of hard solid particles. They are also particularly susceptible to surface degradation by a number of erosion–corrosion mechanisms. Ferrous metals, as we have seen, if left unprotected to natural forces of water and wind, would rust and erode away to nothing as the corrosion products dissolve and wash or are blown away. Contact with corrosive chemicals, contamination with salts and even contact with acid rain can also initiate corrosion and wear when combined with a mechanical action. Metals and alloys chosen for the fabrication of components in industrial production/processing/distribution and marine environments routinely require protection from erosion–corrosion caused by abrasion from slurries, impingement from particles entrained in fast-flowing liquids or gasses, and even impingement from fast-flowing liquids that do not contain particulates.

Erosion–corrosion is a particular issue for the fluid transfer industry where at worst, catastrophic failure, or at best, reduced energy efficiency, results from turbulent flow caused by pitting on internal surfaces of pumps or pipes. Corrosion pits upset smooth liquid flow, creating localised turbulence and localised high-velocity flow rates which leads to even more mechanical damage, further increases in turbulence, and rapidly increasing erosion rates that eventually cause through-wall or joint leaks.

Wet or dry, degradation of a metal or alloy surface or passive oxide layer leads to metal loss or damage and, if left unprotected, ultimately necessitates expensive repair or replacement. In some cases, polymeric materials can be used to repair and provide sacrificial protection for localised wear problems—in certain circumstances erosion-resistant polymeric coatings can prolong the life of equipment indefinitely, depending on whether the problems have arisen from a one-off accidental occurrence or from system design factors.

3.4.1 Essential Chemistry and Technology

Hardfacing, which is the application of a harder or tougher material to a base metal by various welding techniques, is the traditional approach to extending the service life of large, heavy industrial equipment, either for new components or as part of a preventive maintenance programme, where protection against wear and erosion as well as corrosion and high temperatures is required. Thermal spraying techniques like HVOF (high velocity oxygen fuel), flame, plasma and many others, are used to coat specialised structural components with metals, alloys, ceramics and cermet composites to provide protection against extreme wear, erosion and corrosion conditions at high and low service temperatures. These approaches create a hardened layer over a material to provide exceptional levels of resistance to friction and mechanical wear from abrasion and scratching. Plastic linings based on friction-reducing polymers such as PTFE, which are sprayed or dip-coated onto metals provide good levels of resistance to erosion–corrosion, but are not as hard wearing in the most aggressive environments.

Abrasion-resistant coatings for concrete, masonry and metal can also be made by binding hard aggregates, flake or fibres into resin systems exposed to moderately aggressive conditions. Chapter 1.4 described the use of quartz sand with a Mohs hardness of 7 and alumina in the form of calcined bauxite with a Mohs hardness of 9, to give anti-slip/anti-skid properties and to improve the long-term durability of coatings in areas subject to pedestrian and vehicular traffic. Higher hardness ceramic fillers, ranging from common silicon carbide (carborundum) with a Mohs hardness of 9–9.5, to the more exotic, such as boron nitride with a Mohs hardness of 9.5–10 are used to protect metallic equipment that is subject to high mechanical wear forces. Glass flake and ceramic fibres are also used to enhance abrasion resistance by their ability to increase the toughness and the tensile and compressive strengths of polymer composite coatings.

The success of the polymeric coating approach depends as much on the ability of the binder to keep the hard aggregates or fibres locked in, as it does to the inherent resistance to corrosion and abrasion/wear of the base cured resin—almost inevitably the common

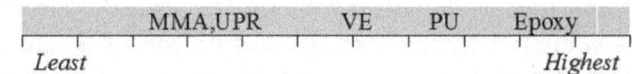

Figure 3.7   Corrosion resistance of erosion-resistant coatings.

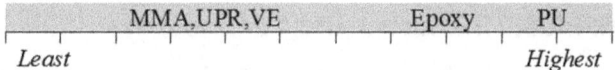

Figure 3.8   Abrasion/wear resistance of erosion-resistant coatings.

commercial types of resin used are epoxy, PU, VE, UPR and MMA, due to their all-round strengths as coating binders. As usual, epoxies and PUs may be solvented, but erosion–corrosion performance is better when used solvent-free. In the same way, MMA, UPR and VE resin performance is contingent on monomer content—the lower the monomer content in the binder, the higher the mechanical per-formance. Figures 3.7 and 3.8 provide qualitative assessments of the resistance to corrosion and abrasion/wear loss for liquid applied erosion-resistant coating technologies.

As a general rule, elastomeric coatings without aggregate fillers are the preferred technologies where resistance to slurries is required; rigid/hard coatings provide sufficient resistance against simple liquids but require high hardness aggregates to upgrade performance when in contact with fluids contaminated with abrasive solids, or in contact with dry abrasive solids.

Aggregate filler particle shape selection has a significant influence on performance of coatings on surfaces subject to particulate abra-sion and erosion: angular ceramic carbides and aluminium oxides resist surface deformation or wear when impacted by coarse abrasive particulates such as pulverised fuel; against fine particulates and where there is a risk of bridging or hold-up of solids on a rough surface then graded ceramic micro-spheres allow tight packing with just sufficient resin to fill the gaps between the spheres to ensure impacting particles only contact the fillers at the surface. Both these approaches limit contact of exposed resin, which will of course be more susceptible to erosion from solid particle impact.

Inclusion of abrasion-reducing non-aggregate fillers can also help improve the wear resistance of polymeric coating materials. These range from friction reducing PTFE and PE powders, to solid lubri-cants such as molybdenum disulfide and finely ground graphite, as well as carbon fibres which not only enhance strength, but also reduce sliding friction resistance.

Exposed aggregate, flake and fibre-filled coatings are applied typically as single layers or are built up to sufficient thickness to provide requisite service life before replacement or patching of localised problem areas is required. Encapsulated aggregate and unfilled coatings are more commonly applied in discrete layers that are characteristically pigmented in different colours to facilitate application (prevent misses) and identify grin-through to prompt maintenance before breakthrough and attack of the "protected" substrate. The discrete layers may be the same coating composition, or may be different but compatible, depending on the severity of the erosion–corrosion environment, but the exposed surface will generally contain higher levels of abrasion-resistant fillers and/or friction-reducing additives.

In highly localised abrasive wear situations, erosion–corrosion resistant coatings can be used as an adhesive to bond preformed abrasion-resistant lining materials for optimum sacrificial protection— these range from basalt and high alumina ceramic bricks, tiles or cylinders to abrasion-resistant steel plate, tungsten and titanium carbide overlay plates, and even thermoplastic linings. As previously indicated in Section 1.9, furan resins would be used as the basis for an adhesive coating where the heat resistance of epoxy, epoxy novolac or VE is insufficient.

There are also proprietary hard, smooth-coating technologies which minimise friction, wear, abrasion and erosion, that are developed specifically for the handling or movement of fluids and gases containing or free from entrained solids.

Application Challenge—Cavitation

Cavitation is a pernicious form of erosion–corrosion involving the formation and collapse of vapour bubbles in moving liquids in contact with metals or other surfaces. It happens in regions of low pressure when changes in velocity and direction of flow in a liquid cause vapour bubbles to form and implode producing high-pressure shock waves which damage susceptible surfaces. Most metals have very good resistance to cavitation but anywhere significant pressure gradients are generated by high-velocity fluid flow over large flat surfaces or within rotating equipment, then cavitation can occur, resulting in surface roughening in the form of sharp-edged craters, which can act as a nucleus for further cavitation bubble formation. Furthermore, the rate at which metals erode and corrode in contact with liquids increases with contacting fluid velocity so in cavitation situations deep pitting of a metal or other surface can occur rapidly.

HVOF-hardened surface-coating composites provide exceptional erosion–corrosion resistance under the highest-intensity cavitation conditions, but are clearly an upfront design solution, not a general on-site repair or maintenance option. Similarly, metal alloys such as nickel–aluminium bronze, although also resistant to high-intensity cavitation, are an upfront design solution where method of fabrication, bimetallic corrosion resistance and other factors permit. UHMWPE (ultra-high-molecular weight polyethylene) coatings, which are sintered directly onto profiled metal, also offer an alternative OEM option.

Two-pack rigid epoxy coatings have proven to be successful in the repair and protection of cast iron, bronze and certain types of cast steel from the effects of erosion and corrosion in applications subject to low-intensity cavitation. They also provide sacrificial protection to medium-intensity cavitation, where even the special bronzes and stainless steels are themselves subject to attack. Two-pack hydrophobic epoxy coatings toughened with rubber modifiers, or flexibilised with elastomeric ur-ethanes, provide higher levels of sacrificial protection, as they are more abrasion-resistant and better able to dissipate shock under medium-intensity cavitation conditions. Two-pack high-strength, heat-resistant elastomeric urethane coatings have been found to absorb the extreme impact pressures from micro-jetting and are proving resilient under higher-intensity cavitation conditions.

Figures 3.9 and 3.10 provide qualitative comparative appraisals of the resistance to cavitation of metals/alloys and erosion-resistant polymers.

3.4.2   Fit-for-purpose Testing

There are a number of conventional and specialist test methods used to predict and validate how coatings, linings and repair composites perform under differing erosive situations. They include the usual ASTM D4541/ISO 4624 pull-off adhesion testing protocols described in Section 3.2, undertaken in conjunction with simulated environmental exposures to evaluate both the initial bonding capability and

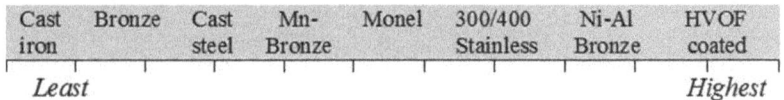

Figure 3.9   Cavitation resistance of metals and alloys.

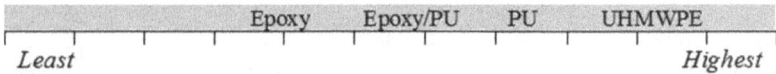

Figure 3.10   Cavitation resistance of erosion-resistant polymeric coatings.

resistance to bondline disruption/susceptibility to permeation under immersion conditions of polymeric coatings. Where appropriate, determination of the resistance to cathodic disbondment is evaluated following the ISO 15711 standard for coatings exposed to seawater or ASTM G8/G42 for pipeline coatings.

ASTM G76 is the test protocol for determining the erosion resistance of hardfaced and thermal spray coated materials subjected to dry solid-particle impingement. Variation of particle hardness, size, velocity and attack angle by this method is a useful means of comparing the resistance of polymeric protective coating and lining materials to particle pull-out and erosion—polymeric coatings that are ductile erode quickly at low impingement angles close to 30° due to cutting and ploughing effects; hardfaced and thermal spray coatings which may be brittle are more susceptible to erosion when impacted at angles of 90° so testing at various impingement angles is very important. ASTM G65 is a method used to determine the resistance of metallic materials to scratching abrasion.

Resistance to erosion–corrosion caused by abrasion from slurries is evaluated through the ASTM D4060 sliding wear test method, and also by the ASTM G6 wet slurry drum test developed to compare the relative abrasion resistance of pipeline protective coatings and linings.

Resistance to erosion–corrosion caused by impingement from liquids without any solid particulate inclusion is evaluated by the ASTM G73 method for solid surfaces in service environments subjected to repeated impacts by liquid drops or jets. For fast-flowing liquids, where vapour bubbles form and implode on solid surfaces, then the ATSM G32 ultrasonic vibratory cavitation erosion test, and tunnel flow method involving water jets in pressure chambers at differing velocities to create a range of cavitation intensities, are used to determine coating wear volume losses over time—the Penn State University Applied Research Laboratory tunnel test facility provides unique cavitation measurement capabilities and correlations with in-service lifetimes for rotating equipment and for flat-surface service.

Recommended Reading

- S. Grainger and J. Blunt, *Engineering Coatings: Design and Application Hardback,* Woodhead Publishing Ltd, UK, 1998, ISBN 10: 1855733692 ISBN 13: 9781855733695.
- J. Lu, Z. Li, X. Gong, J. Han and J. Meng, Resistance to Cavitation Erosion: Material Selection (Chapter 3), *in Wear of*

Advanced Materials, ed. J. Paulo Davim, Wiley, 2013 Online ISBN: 9781118562093 DOI: 10.1002/9781118562093.
• I. M. Kats, *Hydrotec. Constr.*, 1974, 8(6), 539–544 [Investigation of the cavitation resistance of elastic polymer coatings.]

3.5   Chemical-resistant Coatings

Metals and alloys chosen for the fabrication of process vessels, storage tanks, pumps and pipelines for industrial production, processing and transport by pipe or (road, rail or ship) tanker are often susceptible to corrosion and degradation caused by the chemicals handled so are coated or lined with a chemical-resistant barrier to prevent the loss of metal, chemical content, contamination of the environment and cross-contamination of the contents.

Innately, chemical-resistant coatings and linings need to resist a wide range of aggressive processing reagents and solvents in addition to crude oil and gas, produced chemicals, fuels and oils, raw water and industrial effluent including waste. Furthermore, the capability of providing protection in any particular chemical environment depends not only on the nature and physical state of the chemical, but also the concentration, temperature, time of immersion, and any applied stresses. Specifically:

• Polymeric coatings and linings have varying susceptibilities to different types of chemicals and reagents depending upon pH, polarity/solvency and oxidising nature (see Fact File: Chemical resistance of polymeric coatings and linings);
• Non-oxidising gases, or chemical vapours, although highly mobile, are not highly concentrated and, as a result, do not lead to serious chemical attack unless under pressure—liquids are generally more aggressive because the attacking molecules are both mobile and highly concentrated—dry solids rarely cause a problem, but they can do when moistened, as they become extremely aggressive, functioning effectively as high-concentration liquids;
• The concentration of a chemical, particularly in aqueous solutions, is a critical factor—normally the more concentrated the chemical, the more aggressive it is likely to be, but there are cases where coatings may be totally resistant to a concentrated chemical at a low temperature yet disintegrate when exposed to a more dilute concentration of the same chemical at a higher temperature;

- Temperature is probably the most important single factor—as a general rule of thumb the rate of chemical attack can be expected to double for every 10 °C rise in temperature, so whilst there may be no observable attack at 20 °C (68 °F), breakdown may be very rapid at 60 °C (140 °F);
- Where a coating is susceptible to chemical attack, then in most cases, the longer the coating is immersed the more pronounced the attack will be—in some cases there may be no attack for a long period of time but then breakdown may occur very rapidly, in other cases an initial attack may occur but, thereafter, no further attack may take place—intermittent immersion can produce more rapid attack than continuous immersion as, during the period when the coating is not under immersion, the potency of the residual liquid on the surface can increase significantly as a result of evaporation and concentration;
- Abrasion, impact, flexural stresses and pressure applied to a coating when in contact with chemicals can also seriously reduce resistance—degradative attack may accelerate considerably where chemicals are flowing over the coating or lining due to the effects of erosion, increases in temperature due to friction, and a constant supply of fresh attacking media being brought into contact with the coating.

Fact File: Chemical Resistance of Polymeric Coatings and Linings

The aggressiveness of chemicals and reagents towards two-part, high-build polymeric epoxy coatings used as chemical resistance benchmark can be rationalised by type:

- **Alkalis: caustics** such as sodium hydroxide and potassium hydroxide are generally not aggressive, however they can become aggressive at elevated concentrations; **amines and amides** are generally not aggressive, with the exception of ammonia which can be readily absorbed; **alkanolamines** are generally not aggressive, with the exception of MEA (mono ethanolamine) and MDEA (methyl diethanolamine), which can also be readily absorbed;
- **Acids: inorganic/mineral** acids such as sulphuric, hydrochloric, nitric and phosphoric acid vary in aggressiveness depending on their acidity, oxidising potential, concentration and temperature. Hydrofluoric acid which is a solution of hydrogen fluoride (HF) in water is capable of dissolving most materials, especially at high concentrations; **organic** acids such as formic, acetic, acrylic and methacrylic acid are generally very aggressive as they are small molecules, which are typically readily absorbed by polymers;

- **Solvents:** generally attack by swelling polymers and the more polar the solvent, the more likely they will cause problems; **alcohols** are generally not aggressive with the exception of the smaller molecules methanol, MEG (mono ethylene glycol) and bioethanol; **esters and ethers** are generally aggressive, especially at elevated temperatures; **halocarbons** such as dichloromethane, trichloromethane and trichloroethane which are non-polar liquids at ambient temperature are aggressive, especially at slightly elevated temperatures; **ketones** such as acetone (propan-2-one) and MEK (butan-2-one) are highly aggressive, especially at elevated temperatures;
- **Liquid hydrocarbons:** unsaturated **aliphatics** such as pentane, hexane and **cycloaliphatics** such as cyclopentane, cyclohexane and **aromatics** such as benzene and toluene which are non-polar liquids at ambient temperature are generally not aggressive but may be a problem at elevated temperatures and pressures;
- **Gaseous hydrocarbons:** are generally not aggressive, unless wet and likely to from acidic deposits—they can, however, cause physical damage if absorbed into a coating and decompressed rapidly;
- **Mineral Oils**, which are light mixtures of higher alkanes are generally not aggressive;
- **Oxidising agents:** are generally highly aggressive—they include oxygen, ozone, fluorine and chlorine, chlorine dioxide, chlorites, chlorates, perchlorates, hypochlorites, concentrated hydrogen peroxide and other inorganic peroxides, nitric acid and nitrates, sulfuric acid and peroxysulfuric acids.

3.5.1   Essential Chemistry and Technology

The liquid-applied two-part epoxy, VE, PU/polyurea and MMA coatings for concrete containment lining discussed in Section 1.8 are equally capable of protecting ferrous and non-ferrous metals from attack by chemicals, providing they are combined with appropriate anticorrosive primers. There are also many other coating and lining technology options which complement or replace them for site-specific application and service requirements, and also for OEM factory applications where techniques such as vulcanisation, melt processing, powder coating and stove/bake coating are practicable for the internal protection of pipelines, pumps, storage tanks and process vessels.

For each of the different equipment types requiring a chemical-resistant lining, there are usually additional requirements to be factored in when opting for a chemistry and technology. Coatings for internal pipe surfaces for example need to provide protection from the effects of chemical attack, corrosion, erosion and flexure, as well as reducing friction and turbulence in order to maintain or increase flow efficiency. Coatings for the internals of pumps need to protect them from the effects of chemical attack, corrosion, erosion and cavitation, as well as being easy to apply and maintain. Storage tank and process vessel coatings pumps need to protect them not only from the effects of chemical attack, corrosion and erosion but also from movement, impact, washing and cleaning methods, cathodic disbondment, and static discharge, all whilst maintaining the purity of the tank/vessel contents.

Chemically-resistant rubbers, applied either as a coating of unvulcanised material that is then cross-linked and hot bonded in a steam autoclave, or applied as a lining of pre-vulcanised or preformed sheet and cold bonded with adhesive, provide durable and resilient protective linings for cast iron, mild steel, stainless steel and aluminium.

Fluorinated thermoplastic polymers (PTFE, ETFE, ECTFE, FEP, PVDF, PFA), which are favoured for their very low coefficients of friction, abrasion resistance, non-wetting nature and excellent chemical resistance, can be applied by a variety of techniques ranging from melt processing (they melt and flow during baking to form smooth non-porous films), solution casting, aqueous dispersion coating or powder coating. Preformed films can also be resin bonded, a technique used with other high-performance thermoplastics such as PPS and other less frequently used engineering materials with significantly better mechanical and thermal properties than more widely available commodity thermoplastics and thermoset coatings.

Free-flowing solid thermoplastic and certain thermoset polymers can be applied by powder coating, with fusion-bonded epoxy (FBE) powder coatings being chosen extensively for their superior adhesion properties and longevity for pipelines, piping connections and valves operating at moderate temperatures—FBEs are also used to protect concrete reinforcing bars.

Phenol-formaldehyde polymers (phenolics), which have the capability of withstanding low pH and high temperatures, as well as a wide range of chemicals at lower temperatures, along with epoxy-phenolic resins that are used in even more corrosive environments, are both available as liquid-applied coatings. They are low solids and high VOC in nature and rely on solvent thinning for spray application so target dry coating thickness has to be built up from a series of thin wet coats, which require intermediate baking and a final stoving to drive cure through etherification and crosslinking.

The good to excellent chemical resistance shown by solvent-free, liquid applied, two-part coatings formulated from epoxy, VE and PU/polyurea thermoset resins can be boosted by increasing the functionality of the resin and/or hardener components to enhance the ultimate degree and density of crosslinking where needed by finish/ top coats on metal. Raising the core resin functionality typically results in an increase in molecular weight and change in physical form of each resin type, so that solvents or reactive diluents need to be added to fluidise them for ease of application. Care is required, however, as low-molecular-weight resins which rely on gel formation (gelation) and crosslinking in a thermoset process cure with neg- ligible shrinkage, whereas solvented higher-molecular-weight resins cure/dry with some shrinkage and require heated post-/force-cure to prevent against solvent entrapment and the potential for permeation/ exchange/leaching on chemical immersion.

Higher-molecular-weight, high functionality novolac epoxy resins formulated as high-solids solution coatings for ease of application by conventional or airless spray, will gel on application then physically dry when baked to drive off carrier solvent—heating is essential to relieve shrinkage stress build, trapped solvent and eliminate air voids in an annealing process before immersion service. If vitrification can be avoided before full crosslinking, very high levels of resistance to chemicals and heat can be achieved. The alternative option of using low-viscosity, multi-functional reactive diluents, rather than solvents, works particularly well with low- to moderate-molecular-weight epoxy novolacs, which gel and cure with negligible shrinkage and no-voids at ambient temperatures and complete their cross-linking through postcure in service—no force cure is required. The use of water as a dispersive carrier is not a viable option where chemical resistance is required as, although aqueous emulsions dry with minimal volume shrinkage, they result in microporous (breathable) coatings, which are not resistant to osmotic permeation (see Section 3.6).

There are a number of proprietary approaches to the production of highly crosslinked, yet flexible polymer coatings to maximise re- sistance to chemicals. These include bimodal epoxy technology based on blends of low molecular weight and high molecular weight resins, and also siloxane hybrids with epoxy and novolac epoxy resins—here inorganic backbone polymer linkages ($-SiO_2-$) impart resistance to oxidation and heat, although coatings require a post- cure before immersion service when solvent thinning is used for application.

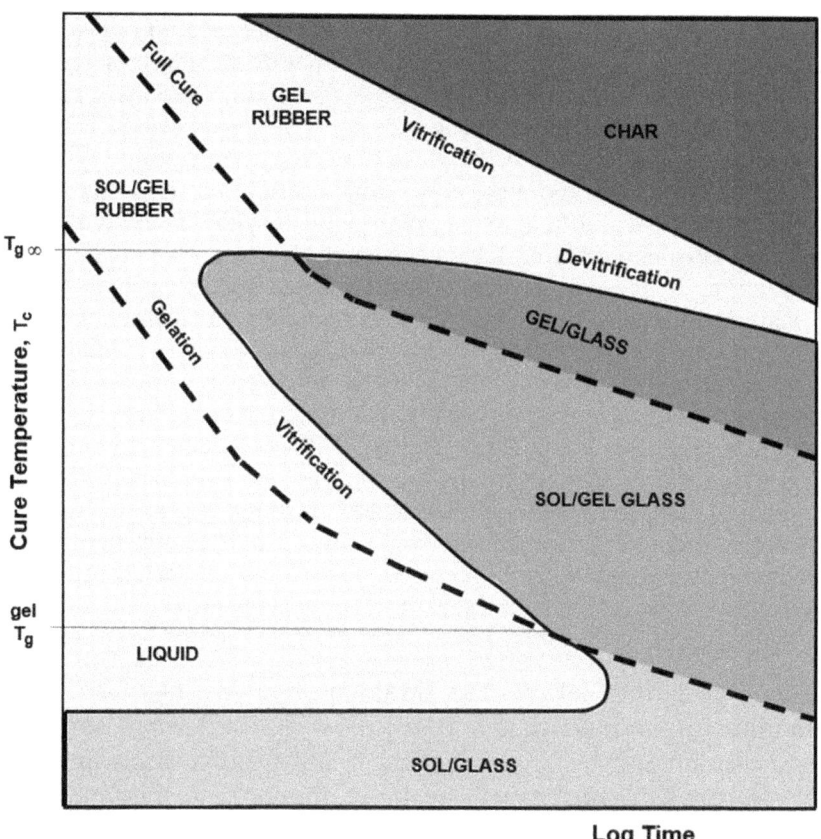

Figure 3.11   Time–temperature-transformation (TTT) diagram for thermo-
set polymers.
Reproduced from ref. 3 with permission from John Wiley and
Sons. Copyright © 1983 John Wiley & Sons, Inc.

Fact File: Thermoset Polymer Gelation, Crosslinking and Vitrification

Thermosetting polymers are created from liquid resins, monomers and
curing agents and/or catalysts through reactions involving chain extension
and branching, leading to an increase in viscosity in the liquid (sol) stage
as expressed in a Gillham time–temperature-transformation (TTT) diagram reproduced in Figure 3.11, where the time to gelation and vitrification is plotted as a function of isothermal cure temperature. At a
particular conversion of functional groups, viscosity increases infinitely
and an insoluble crosslinked rubbery polymer gel fraction (sol/gel) forms
in the reaction mixture in the presence of unreacted or partially reacted
monomers which may still be soluble (sol fraction). The Carothers or,
more recent Flory-Stockmayer, equations can be used to predict the degree
of   conversion   for   gel   point   from   the   functionality   and   relative

concentration of the ingredients. As reaction proceeds and at near full conversion of functional groups in a stoichiometric system the sol fraction disappears with the creation of a solid (gel/glass) thermoset polymer macromolecule or interpenetrating network of macromolecules. The onset of vitrification occurs when the glass transition temperature (T_g) of the curing polymer rises above the cure temperature when the system transforms from a gel into a glass at which point polymer mobility reduces, reaction rates change from kinetically controlled to diffusion controlled, and only an increase in the cure temperature can restart reactions for any remaining unreacted functionality. Postcure makes sure all monomers are built into the polymer network to ensure no soluble fraction remains and also ensures further crosslinking of the glass to maximise crosslink density. When solvented high-molecular-weight resins vitrify following solvent drying, it is possible for the sol/gel stage to be missed altogether and unreacted functional groups are then unable to move or get close enough to complete reaction unless heated to soften and anneal together to permit completion of crosslinking—this is particularly detrimental where chemical resistance is required, as residual unreacted functional groups act as target sites for attack in incompletely cured coatings.

When designing solvent-free, liquid-applied coatings for chemical resistance, it is clear that a balance is required when increasing crosslink density so as to avoid the risk of vitrification and other adverse impacts—increasing the functionality of resins and use of multifunctional diluents and/or curing agents shortens pot-life and workability of mixes and can also reduce key coating mechanical properties through embrittlement. There are, however, molecular fortifiers, also known as antiplasticisers, available as liquids or low-melting-point solids, which can be used as additives to improve the modulus and yield stress of an epoxy functional thermoset by filling free volume and, through specific physical interactions which can be polar in nature or involve covalent bonding, to increase the effective crosslink density of the network. More commonly, most chemical-resistant coatings for metals and alloys incorporate barrier pigments and fillers such as micro-glass flake to maximise resistance to permeation and chemical attack.

Application Challenge—Static Dissipation

Where there is a potential explosion risk due to static build-up, discharge and sparks within fuel, solvent or other flammable chemical storage tanks, it is vital that chemically resistant lining materials are conductive or static-dissipative in nature, to prevent operating hazards. As previously discussed in Section 1.4, synthetic resin coatings are inherently insulating in nature (electrical resistance $>10^{12}$ Ω) and, in practice, most types of chemical-resistant coating can be modified to include electrically conductive fillers to make them static-dissipative (surface resistance $>10^5$ Ω but less than 10^{12} Ω) to meet the requirements of the many different standards around the world governing electrostatic discharge (ESD) control. So-called static-dissipative coatings act as conductive coatings when used over a conductive primer or an earthed metal substrate; they are also used directly on insulating surfaces such as GRP tanks and silos.

3.5.2 Fit-for-purpose Testing

As indicated above, there will usually be a number of requirements over and above chemical resistance that need to be considered when testing the fitness for purpose of coatings applied in different ways for the protection of various different equipment types from chemical attack. In terms of resistance to chemical reagents, however, there are a number of fundamental methods: ASTM C868 governing testing of coatings for resistance to corrosive attack by chemicals under constant immersion; ASTM D543 covering the evaluation of all plastic materials including cast, hot-moulded, cold-moulded, laminated resinous products, and sheet materials; and specific methods such as ASTM G20 for assessing pipeline coatings, and ASTM D6943/NACE TM0174 standard methods for appraising coatings exposed to cold wall effects and decompression.

As usual, there can be significant variations in performance against any given chemical between the same generic type of polymeric coating or lining depending on how they are formulated and applied—usually the easier they are to apply, the lower the resistance to chemicals amongst other properties—and there can be differences on exposure to any given chemical depending on the concentration, temperature, time of contact and applied stresses. However, it is possible to make a simplified generic comparison of the chemical resistance of polymeric coatings and linings as presented in Figures 3.12–3.14 for chemical-resistant coatings formulated from key resin technologies.

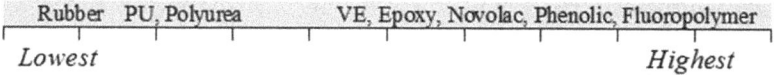

Figure 3.12 Aqueous acid resistance of chemical-resistant coating technologies.

Fluoropolymer, PU, Polyurea	VE, Epoxy, Novolac, Phenolic, Rubber
Lowest	*Highest*

Figure 3.13 Aqueous alkali resistance of chemical-resistant coating technologies.

Rubber, Fluoropolymer	PU, Polyurea, VE	Epoxy, Novolac, Phenolic
Lowest		*Highest*

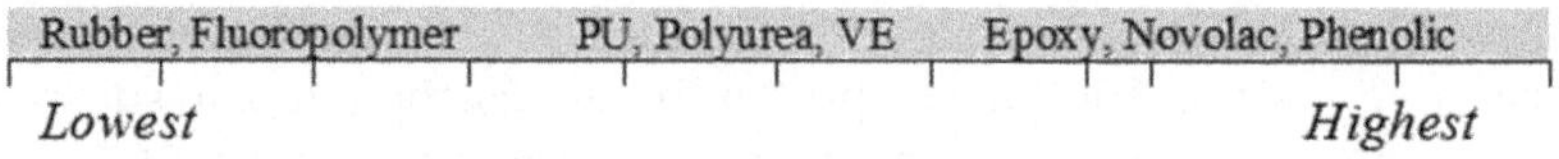

Figure 3.14 Solvent/hydrocarbon resistance of chemical-resistant coating technologies.

Recommended Reading

- C. M. Hansen, *Hansen Solubility Parameters: A User's Handbook*, CRC Press, 2007, ISBN 0-8493-7248-8.
- J. B. Enns and J. K. Gillham, *J. Appl. Polym. Sci.*, 1983, **28** (8), 2567–2591.
- J. K. Gillham, *Encyclopaedia of Polymer Science and Engineering*, John Wiley, New York, 2nd edn, 1986, pp. 519–524.
- *EI 1541 Performance Requirements for Protective Coating Systems Used in Aviation Fuel Storage Tanks and Piping*, Energy Institute Publications Online, UK, 2009, ISBN: 9780852935668.
- *'Performance Standard for Protective Coatings for Cargo Oil Tanks of Crude Oil Tankers'*, International Maritime Organisation Resolution MSC. 288(87).
- D. Stauffer, A. Coniglio and M. Adam, *Gelation and critical phenomena*, Adv. Polym. Sci., 1982, **44**, 103.

3.6   High-temperature Coatings

There are many instances where coatings for metal and concrete need to provide protection from corrosion, erosion and/or chemical attack at high operating temperatures. As a general rule, conventional two-part thermoset polymeric materials are suitable for use in dry conditions at temperatures up to twice their HDT (see box), although in wet conditions their ability to survive is frequently restricted to temperatures below 120 °C (~250 °F). In oxygen-rich environments, organic polymers are vulnerable to thermo-oxidative degradation at temperatures over 250 °C (~500 °F), as well as to pyrolysis and thermo-chemical decomposition in the absence of oxygen as temperatures rise above 300 °C (~600 °F) when crosslinked network and linear C–C bonds in the polymer break down through homolytic and heterolytic cleavage. Introduction of inorganic silicon backbone polymer functionality improves resistance to oxidation and resistance to heat as the Si–O bond is already oxidised and has a significantly higher bond strength (108 K cal mole^{-1}) compared to a C–C bond (83 K cal mole^{-1}).

Fact File: Heat Resistance of Thermoset Polymers

When thermoset polymers are heated, a number of things can occur as a result of thermal motion of the polymer segments increasing as the temperature rises. When the temperature reaches the glass transition point (T_g), polymers transition from a hard, glassy material to a soft, rubbery material. T_g can be determined by several different methods including ISO 11357-2 differential scanning calorimetry (DSC), differential thermal analysis (DTA), ISO 6721-11 dynamic mechanical analysis (DMA), and also by dielectric spectroscopy or dilatometry—the measured value for T_g varies somewhat, depending on the equipment and method selected. DMA can also be used to monitor change in mechanical modulus as the temperature rises, as can heat distortion temperature (HDT) determined in accordance with the ISO 75/ASTM D648 standards, which are commonly used to benchmark heat resistance for polymer composite coatings where engineered filler and fibre interactions with the polymers lead to an increase in modulus over and above the T_g of the polymeric component—HDT determines the temperature of deflection under load (essentially flexural stress under three-point loading) with different types of test specimen and different constant loads being defined to suit different types of material. ASTM D1525 Vicat softening point can also be used for polymeric materials that have no explicit melting point.

As thermoset polymers are exposed to heat within their heat-resistance capabilities, it is possible for secondary reactions to occur within the thermosetting polymer, as well as with fillers and fibres, leading to an increase in mechanical properties over and above initial cure capabilities. Heat resistance, determined from T_g or HDT, can improve significantly by the process often referred to as postcuring. This is in contrast to thermoplastic polymers which under the influence of heat simply soften and lose mechanical integrity as they approach their T_g or melting point; providing temperature exposure is not excessive and exposure is below thermo-oxidative/-chemical margins, thermoplastic polymers will, on cooling, return to normal with no adverse effect on mechanical properties, although a change of shape or form may have taken place if unsupported when heated.

When thermoset polymeric coating materials are exposed to temperatures over and above their wet heat-resistance capabilities, oscillation and vibration of the polymer segments tends to increase dramatically, forming free volume into which water or other individual liquid molecules may move and take a random walk through the polymeric coating from the surface through to the coating/metal interface. Liquids will of course migrate through any pores, channels or capillaries in coatings formed as a result of flaws from air entrainment or incomplete wetting of pigment/filler particles with binder during manufacture or mixing, or from volatile solvent evaporation during curing, so increases in free volume on heating only aggravate inherent porosity and permeability issues.

Heat-resistant polymeric coatings for immersion service can also be sensitive to diffusion and osmotic flow. With thermo-osmosis any significant thermal gradient across a coating can result in transport of heat and liquids through a coating unless potential cold wall effects are lessened by insulation. Heat-resistant polymer coatings for aqueous immersion service are also particularly susceptible to osmotic pressure differences as water can move through to the coating/metal interface as the result of diffusion under a concentration gradient unless substrate salt contamination has been removed prior to coating.

3.6.1 Essential Chemistry and Technology

The intended application usually dictates the polymeric resin binder choice, whereas filler selection invariably involves powdered metals, mica, silica, alumina and other ceramics—all of which help maximise heat resistance by either: decreasing the thermal coefficient of expansion; increasing thermal conductivity; increasing thermal shock resistance; and with, or without, the aid of coupling agents by bonding to restrict thermal motion of polymer segments at temperatures at or above their intrinsic softening points.

The highest heat-resistant coating applications for chimney stacks, hot duct work/piping and process equipment calls for silicone resins and initial bake temperatures of at least 150 °C (~300 °F) to convert them to polysiloxane polymers. On exposure to temperatures above 300 °C (~600 °F), the organic components pyrolytically decompose, leaving crystalline silicate polymers of stoichiometric composition $(-SiO_2-)_n$ with fillers and pigments firmly bound within, and to, the coated substrate. This results in corrosion protection at temperatures of up to 760 °C (~1400 °F), although adhesion and flexibility/impact resistance all diminish on ageing.

However, not all heat-resistant coatings incorporating silicone need to be pyrolysed to form crystalline silicates to be useful—methyl-substituted polysiloxanes are water repellant and relatively hard and resist temperatures up to around 200 °C (~400 °F); phenyl substituted polysiloxanes are resistant to temperatures of up to 250 °C (~500 °F). Cold blending or copolymerisation of silicone resins with conventional binder resins such as alkyd, acrylic, UPR, VE and epoxy leads to coatings with enhanced adhesion, mechanical properties and dry heat temperature capabilities up to around 220 °C (~430 °F). Epoxy novolac resin (EPN) blends with epoxy functionalised polysiloxanes are reported to resist thermal shock up to 260 °C (~500 °F)—for immersion service, however, a post-cure is required to remove the solvent added at point of use to facilitate their application.

High wet and dry heat resistance in pure organic polymer systems has by tradition been achieved through the use of components with high aromatic character as well as through rigid crosslinks. Phenolics (epoxies cured with phenolic resins and polyamine crosslinkers) are used in severely corrosive environments where sustained immersion is required such as tank lining and heat transfer equipment—challengingly, they have a significant demand for solvent to permit application and the ultimate dry coating thickness has to be built up from a series of thin wet coats with intermediate bakes, and a final bake before immersion service is feasible.

Epoxy novolac resins (EPNs) combined with multifunctional polyamine curing agents are also naturally tolerant of elevated temperatures in their own right—their high inherent aromatic character and high crosslink density capability gives much lower average molecular weights between crosslinks (M_c) than other thermoset resin types.

Fact File: Average Molecular Weight Between Crosslinks (M_c) of Thermoset Polymers

The average molecular weight between crosslinks (M_c), also defined as the total sample weight that contains one mole of effective network chains, is determined typically by dynamic mechanical analysis from the shear storage modulus (G') of the cured network in the rubbery region at temperatures well above T_g through the Prime equation: $M_c = \rho RT / G'$; where ρ is the density of the cured material, R is the gas constant and T is the temperature in Kelvin.

Incorporation of covalently bonded silicon or other inorganic fillers and fibres, interpenetrating polymer networks, and/or free-volume fortifiers also helps boost wet heat performance of EPNs for continuous use at high temperatures, and for shorter term exposures to pressurised steam such as experienced during cleaning/dewaxing hydrocarbon processing equipment.

3.6.2 Fit-for-purpose Testing

ASTM D2485 is the standard embracing key methods for evaluating coating systems designed for interior service, and for coatings designed for exterior (weather-exposed) service, on steel for chimneys, heat exchangers, reactors, boilers, evaporators, pipes or autoclaves operating at temperatures beyond 200 °C (~400 °F). Good adhesion retention and the absence of cracking/chalking are the key indicators of heat resistance and stability from thermal cycling and prohesion corrosion testing following incrementally increasing heat treatments up to 400 °C (~750 °F). ASTM D5499 is the standard covering testing for the validation of the heat resistance of polymeric linings for flue gas desulfurisation systems with typical maximum operating zone temperatures of 200 °F (93 °C) and 350 °F (177 °C).

There are a number of industry-specific test methods used to validate the capabilities of internal linings for process vessels, tanks and pipelines containing chemicals and hydrocarbons operating at high temperatures and pressures – they are the NACE TM0174 atmospheric pressure and pressurised Atlas Cell test which permits evaluation of the influence of cold wall gradients/potential for cold wall blistering with and without lagging options, and the NACE TM0185 three-phase autoclave test for liquid, hydrocarbon and gas exposure. In conjunction with ISO 16773 Electrochemical Impedance Spectroscopy measurement of coating impedance retention after exposure in liquid, vapour and gas phases, these methods provide useful correlations with wet heat/immersion service performance capabilities as discussed previously in Section 3.3. Figure 3.15 provides a qualitative assessment of the resistance to wet heat/immersion for the various high-temperature coating technologies.

The heat resistance capabilities in dry conditions of polymeric coating materials is often determined using the ASTM D648/ISO 75 methods for heat distortion temperature (HDT)—also known as heat deflection temperature or deflection temperature under load (DTUL). Heat resistance in dry air can also be determined by the ISO 11357-2

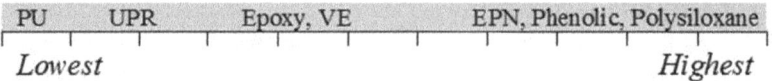

Figure 3.15 Wet heat/immersion resistance of high temperature coatings.

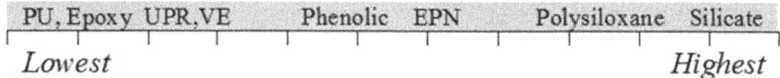

Figure 3.16 Dry heat resistance of high temperature coatings.

and ISO 6721-11 methods for glass transition temperature (T_g). Figure 3.16 presents a qualitative assessment of the resistance to dry heat for high-temperature coating technologies.

Thermo gravimetric analysis (TGA) is a useful means of checking for weight changes from unreacted or unpredicted volatile ingredient losses at curing (including post-curing) temperatures. TGA can also be used to study the formation of gaseous degradation products like carbon monoxide, carbon dioxide and water vapour on exposure to air at significantly higher temperatures.

Application Challenge – Explosive Decompression

High temperature linings/coatings used for the protection and/or restoration of process vessels and equipment in the oil and gas recovery and processing industry are frequently exposed to upset conditions. These range from unexpected temperature excursions over and above design temperature resistance capabilities, to unplanned hot decompressions/blow downs from high operational pressures. In the latter case, gases such as water vapour, methane, carbon dioxide and hydrogen sulphide trapped in the coatings create pressures which are considerably greater than in the free gas phase and any voids and weak points are then subjected to extreme pressure and can lead to catastrophic (explosive) failure of the coating. Under routine pressurised operating conditions, these gases absorb through the surface of the polymeric lining which becomes saturated in the gas, oil and water phases. Furthermore, gas diffuses throughout the lining at a rate that is dependent upon the pressure of the gas and the diffusion coefficient of the gas through the coating, and when saturated exerts a pressure equivalent to that in the gas phase, so in instances when the pressure of the gas phase reduces rapidly, trapped gas cannot diffuse out of the coating quickly enough to relieve the pressure differential. Under normal operating procedures, equipment is permitted to decompress in a controlled manner, which permits trapped gases to diffuse out and pressures equilibrate harmlessly. Design options for high temperature/pressure immersion coatings fall into two categories: thin coatings, which allow liquids and gases to penetrate and escape quickly; or thick coatings which provide a barrier to permeation and limit thermoosmosis and diffusion effects. With both approaches the key to success involves maximising the adhesion of the coating material to the substrate and minimising the number of voids within the coating.

Recommended Reading

- J. R. Kosek, J. N. DuPont and A. R. Marder, *Effect of Porosity on the Resistance of Epoxy Coatings to Cold-Wall Blistering, Corrosion,* 1995, 861.
- R. B. Prime, *Thermal Characterization of Polymeric Materials,* ed. E. A. Turi, Academic Press, New York, N.Y., 2nd edn, 1997, pp. 1379–1766.

3.7   Bonding, Rebuilding and Repair

In addition to providing protection for metal from corrosion, erosion and chemical attack over a wide temperature range, polymeric materials are also used to solve many diverse construction, rebuilding, repair and mechanical engineering problems. These vary from structural bonding, to rebuilding corroded/eroded metal prior to application of a protective coating, through to emergency and permanent repair of holes and cracks, to pourable grouting/irregular shimming, and ultimately to compliant composite restoration and repairs that return strength to metallic substrates with thinned-wall defects or through-wall defects.

3.7.1   Essential Chemistry and Technology

Two-part epoxies have been known since the early 1950s as structural adhesives for metal due to their high adhesion, high strength, high load bearing and cold bonding capabilities; nowadays they are used universally as a replacement for traditional mechanical engineering joining methods. Other frequently used polymer technologies for structural bonding are: one-/two-part PUs, which are recognised for their toughness and flexibility at low temperatures; two-part MMAs, for their high tensile, shear and peel strengths and resistance to impact across a wide temperature range; two-part and two-step acrylics, known for their high shear and peel strengths, good flexibility and impact resistance; one-part cyanoacrylates, for bonding small areas and closely fitting surfaces; one-part anaerobics, for closely fitting surfaces such as thread fastening; and hot-melts, for low stress product assembly. Figure 3.17

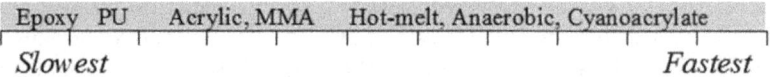

Figure 3.17   Speed of cure of resin and polymer adhesives technologies.

presents a qualitative assessment of the relative speed of cure for the key resin and polymer adhesive technologies.

Fact File: Anaerobic Adhesives and Sealants

Anaerobic adhesives and sealants are designed to remain liquid until isolated from oxygen, and then cure on intimate contact with metal/metal ions, or other activators where non-metallic substrates are involved. Based primarily on methacrylate esters, they harden to form strong bonds and seals between metal surfaces of threads, joints, fastenings, retaining bearings. Other vinyl polymerisable monomers containing appropriate polymeric modifiers are used for bonding dissimilar substrates such as metal to elastomers, wood or composites. Heat, primers and cure accelerators are all used to promote the hardening process.

Structural adhesives intended for automated mixing and dispensing, especially by two-part cartridge guns with static mixer nozzles, generally are formulated without fillers. Where manual mixing is possible, functional fillers are included at filler to binder ratios between $0:1$ and $2:1$ depending on the filler purpose and application method: low filler contents for pouring, spraying or brushing; higher filler content pastes requiring use of a spatula, trowel or mastic pressure gun. Inorganic pigmentation can be added to both parts (each a different colour) of two-part formulations as a visual means of verifying complete mixing. Electrically conductive fillers such as powdered copper, silver and graphite, or thermally conductive metal/alloy fillers are typically added to provide polymeric adhesives with specific conductance properties. Figure 3.18 presents a quantitative summary of electrical conductivity of various materials based on volume resistivity measurements[†].

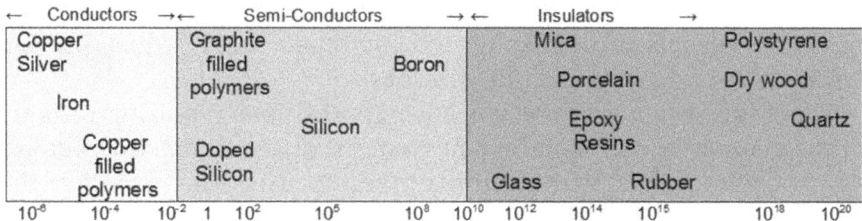

Figure 3.18 Volume Resistivity (ohm cm) of selected metals, materials and polymers.

[†]Redrawn based on data taken from http://chemistry.about.com/od/moleculescompounds/a/Table-Of-Electrical-Resistivity-And-Conductivity.htm

Two-part epoxies are also used extensively for the rebuilding, re-surfacing and repair of holed, cracked and worn machinery and equipment – their adhesion to metals such as stainless steel, carbon steel, aluminium, brass and copper, as well as natural and synthetic materials, is unsurpassed. Other thermoset polymer technologies used to formulate pastes or putties for metal refurbishment include UPR, PU, cyanoacrylate/epoxy hybrids, and EPN; VE based repair pastes are used generally for marine filling and fairing applications above and below the waterline where adhesion to fibre reinforced plastic (FRP) as well as aluminium is required.

Rebuilding and repair grades are filled generally with powdered metals, alloys, ceramics and low oil absorption pigments and particulate mineral fillers and extenders which can all be used at high loadings without adversely increasing viscosity or thixotropy—in practice, significant levels of thixotropes need to be added deliberately to prevent heavy filler settlement during storage/transportation and for controlling slump after mixing and application. Alloys of iron with phosphorus, which are non-magnetic, and alloys of iron with silicon, which are ferromagnetic are used to impart metallic properties to polymers such as appearance and thermal conductivity without compromising inherent corrosion resistance. Aluminium flake and powdered steel and titanium, among other common metals, find particular use in two-part epoxy repair putty sticks which are favoured for filling cracks and gaps or sealing leaks prior to application of a more permanent seal. Two-part repair pastes and putties are relatively easy to mix and apply without the need for specialist tools—they are based on simple volume mix ratios with putty stick variants having their ingredients premeasured and amalgamated into a single piece with a filled resin component on the outside protecting the filled curing agent on the inside. Most grades can be machined after cure using conventional tools, and there are a number of distinct proprietary methodologies used to make resurfacing repair pastes, which provide low-friction and self-lubricating finishes.

There is a wider range of specialist applications requiring variants of the two-part epoxy metal repair pastes that are sufficiently fluid to pour or pump/inject with minimal pressure. These are used for: the protection of backing wear plates in cone crushers and grinding mills; locking the inner and outer eccentric bushings in cone crushers; casting-in-place permanent engine and machinery supports; injection of bearing seats or column bedding; casting of components and formation of irregular shims; and, bonding doubler plates/stiffeners on corroded decks/beams. These normally contain the same fillers as the repair pastes, but exclude thixotropes as they need to be free-flowing, with binders based on unplasticised low viscosity epoxy resins combined with low viscosity highly reactive curing agents which cure rapidly and develop permanent high compressive strength support, or on low viscosity PU resins where flexible and tough backing materials are needed.

Two-part repair pastes, putties and resins in combination with metal doubler plates and/or synthetic fibres and woven fabrics have for many years provided a recognised field solution for the repair of weakened or holed vessels and pipework where moderate pressure retention capabilities as well as corrosion protection are required. Where equipment has already failed, and been taken off-line, then the same filled two-part PU, UPR and epoxy repair grades mentioned previously are used to embed reinforcement fabric or metal plate into place to protect and strengthen damaged areas—the more effectively the damaged area can be cleaned of oil or grease and abraded (preferably by blasting), the sounder the mechanical key and adhesion possible, the more permanent the repair solution.

Emergency sealing of low pressure leaks from fractures and cracks in metal piping and equipment is possible with fibre and fabric tapes, bandages or patches pre-impregnated with resins which range from petrolatum wax, to water-dip-activated PU, to heat-activated epoxy, to sprayed-initiator-activated or light–activated MMA, to sunlight-/UV-curable UPR or VE. On-line emergency repairs made with these technologies can be permanent depending upon operational pressures and level of corrosion associated with the failure.

Codal compliant repair of thinned wall defects or through wall defects on straight pipes, complex geometries, tanks and vessels is governed by the ISO 24817 and ASME PCC-2 International Standards which classify all intended repairs based on temperature/pressure/content service usage, and by lifetime—short lifetimes of up to 2 years, where the repair is required to survive until the next shutdown, after which equipment shall be replaced; long lifetimes up to 20 years where the repair is required to reinstate equipment to its original design lifetime or to extend its life. Uni-directional, bi-directional, and multi-directional woven E-glass, carbon fibre and Kevlar fibre reinforcements are all used in combination with epoxy, UPR, VE, MMA and PU resins which are applied either as composite wrap systems, pre-impregnated wraps, or as adhesively bonded pre-cured wrap systems—see Section 4.3 for more detail about thermoset and UV-activated prepreg technologies.

3.7.2 Fit-for-purpose Adhesion Testing

There are a number of important general test methods used to validate the integrity of structural adhesives and repair materials based on joint design or other application specific requirements.

Structural adhesives for rigid metals are formulated typically with high shear, tensile and compressive strength as joints are engineered usually to have as little exposure as possible to cleavage, peel or torsion stresses—in practice, joints are also designed wherever possible to ensure applied loads are transferred by shear or compression so that direct or induced tensile stresses are minimised. BS EN 1465/BS 5350 Part C5/ASTM D1002/ASTM D3166 describe methods for the measurement of shear adhesion of a multitude of shear test geometries for lap joints (single, double and modified) amongst other configurations, but the single-lap shear test is the most widely used method for verifying the tensile shear strength of adhesively bonded joints. Where tensile butt joints need to be used, BS EN 15870/ASTM D897/ASTM D2095 are used to evaluate the load transfer capabilities under direct tension. For rigid joints subject to cleavage stress from the application of a tensile force at one edge of a joint, then the BS 5350: Part C1/ASTM D1062 compact tension test method is useful for assessing compliance of an adhesive under tension and cleavage. Adhesive resistance to cleavage from tensile stresses applied at the crack tip of flat-bonded configurations is determined in the ASTM D3762 wedge cleavage test method which is more severe than lap-shear or peel tests—ISO 8510: Part 2/ISO 11339/ ASTM D1876 describe T-peel adhesion methods for use with thin flexible metal adherends.

For metal rebuilding, resurfacing and composite repair materials, ASTM D4541 pull-off testing is used to determine adhesive strength as well as for checking mode of failure of multi-layer systems. Figure 3.19 presents a qualitative assessment of the adhesive strength of structural adhesive resins.

3.7.3 Fit-for-purpose Mechanical and Physical Testing

The critical mechanical properties required to validate the structural integrity of general rebuilding and repair materials are

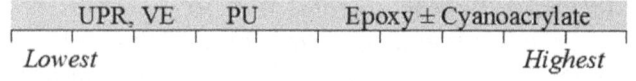

Figure 3.19 Adhesion to steel of structural adhesive resins.

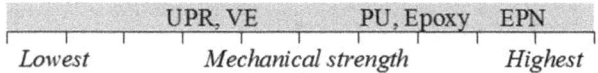

Figure 3.20 Mechanical strength of rebuilding and repair materials for metal.

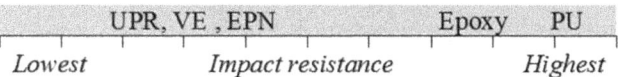

Figure 3.21 Impact resistance of rebuilding and repair materials for metal.

ASTM D638/ISO 527-1 tensile strength/modulus, ASTM D790 flexural strength/modulus and ASTM D695/ISO 604 compressive strength/modulus. Figure 3.20 presents a qualitative assessment of the relative mechanical strength of metal rebuilding and repair technologies.

The ISO 24817/ASME PCC-2 standards for compliant composite repair call for data from a number of specific mechanical and physical tests: ASTM D3039 Young's modulus and Poisson's ratio; ASTM D5379 shear modulus; ASTM D790 flexural modulus; ASTM D2240/ISO 868 Shore hardness; ASTM E831 or ISO 11359-2 thermal expansion coefficient and ASTM D648/ISO 75 heat distortion temperature. This data, along with failure pressure limits established from practical testing as defined in Annex D of ISO 24817 provides confidence limits of energy release rates needed in repair design calculations of thickness, number of layers of reinforcement and length depending on defect and class types.

Other physical properties that help confirm the robustness of metal composite rebuilds and repairs in service include ASTM D256 Izod impact resistance, and Figure 3.21 provides a qualitative assessment of the relative impact resistance characteristics of the key metal rebuilding and repair technologies.

It is also usual for ASTM F433 thermal conductivity measurements to be made to validate heat transfer capabilities, along with ASTM D257/IEC 60093 volume resistivity and surface resistivity measurements made to verify electrical insulation properties for specific applications.

Recommended Reading

- *ISO 24817 Composite Repairs for Pipework: Qualification and Design, Installation, Testing and Inspection,* International Organization for Standardization, Switzerland, 2015, ICS: 75.180.20.

- *PCC-2 Repair of Pressure Equipment and Piping*, American Society of Mechanical Engineers, 2015 ISBN: 9780791869598.
- E. Zamzam, *Repair Of Damaged Metal Pipes Using Composite Materials,* Lambert Academic Publishing, UK, 2015, ISBN-10: 3659759759; ISBN-13: 978-3659759758.
- *ASTM D5363, Standard Specification for Anaerobic Single-Component Adhesives (AN),* ASTM International, West Conshohocken, PA, 2008.

3.8 Under Insulation and Fireproofing Corrosion Control

Corrosion occurring underneath thermal insulation or protective fireproofing applied to structural steel can be very difficult to detect, and the consequences if not identified and controlled can be expensive and disastrous in cases where there is loss of containment or integrity. Prevention of premature failure from corrosion under insulation (CUI) and corrosion under fireproofing (CUF) of steel operating equipment and piping is required at both high and cryogenic temperatures—anywhere in fact insulation gets used to protect personnel from the risk of injury from hot or cold surfaces and minimise heat loss or heat gain for efficiency purposes.

Fact File: Contributing Factors to Corrosion Under Insulation and Fireproofing Materials

The risk of CUI and CUF on carbon steel exists between $-4\ °C/25\ °F$ and $+175\ °C/350\ °F$, increasing between $+60\ °C/140\ °F$ and $+120\ °C/250\ °F$ when general and localised corrosion thrives where moisture, oxygen and salts collect under insulation or fireproofing. Moisture can get trapped during installation of insulation, and can also infiltrate as rain, fire water or process liquid through leaking/damaged insulation or by condensation from the atmosphere during plant shutdown periods. Normally the amount of dissolved oxygen decreases as the temperature rises, but under insulation there is a closed system poultice effect, which maintains sufficient levels required by the corrosion process. Salts required for electrolyte formation, can be present from contaminants from the original or decomposing insulation as well as from atmospheric pollutant or other leakages through damaged insulation. It seems that the metal temperature affects the corrosion process in different ways: at low temperatures corrosion rates are suppressed, but increase exponentially as the temperature increases following an Arrhenius relationship; at temperatures above 120 °C/250 °F, however, any trapped moisture tends to evaporate drying out insulation and reducing wet time—salt contaminants also concentrate during the evaporation process which during shutdowns readily reabsorb returning moisture and migrate to the cooled metal interface to accelerate the corrosion process. Metal temperature can also negatively impact the corrosion process by reducing the life of any applied coatings and insulation.

Austentitic stainless steels are susceptible to both localised crevice and pitting corrosion under insulation and fireproofing. There is also a high risk of stress corrosion cracking (SCC) which is heightened in the presence of chloride salts at temperatures above 60 °C/140 °F, in the presence of moisture, and where there are residual or applied stresses.

Whenever CUI or CUF is a concern, a corrosion-resistant protective coating can be applied directly to the steel before insulation or a fireproofing system is installed to mitigate the risk. Where insulation is applied, especially types requiring cladding to keep them in place, encapsulation with a seamless and flexible coating provides further protection. This section describes both corrosion-resistant coating and encapsulation systems that are suitable for new installations, for planned or unplanned offline remediation, and also for application whilst equipment and plant are on-line.

3.8.1 Essential Chemistry and Technology

The prevention of corrosion under both insulation and fireproofing requires suitable anti-corrosive coating systems to protect the underlying steel. There are different approaches depending upon whether the steel can be coated before installation, if the equipment can be taken offline to permit replacement of failed previously applied coatings, and also on the level of remediation required for any corrosion which may have been detected. There are also different temperature capability options dependent upon intended service conditions.

Where steel can be blast cleaned to ISO 8501 standard Sa $2\frac{1}{2}$/SSPC-SP10 minimum, or mechanically prepared by chipping, scraping and wire brushing to remove loose millscale and rust to ISO 8501 St 3/SSPC-SP3 minimum, the various conventional solvent-based and solvent-free polymeric corrosion resistant coatings described in Section 3.3, in conjunction where necessary with the primers discussed in Section 3.2, are often suitable. As their serviceable lives are shorter generally than the expected operational life of equipment, they do need regular inspection and reapplication/refreshing to maintain integrity and mitigate ongoing risk of CUI and CUF. The various high temperature coatings reviewed in Section 3.6 provide long term protection on steel from which mill scale, rust and old coatings has been removed by abrasive blasting to ISO 8501 St 3/SSPC-SP3 minimum.

Initially, naturally occurring oil based coatings were used directly under lagging but have now been surpassed generally by solvent-based asphaltic mineral fibre mastic tapes which combine fibrous insulation with an organic binder to bond them in place and to provide seamless protection. Synthetic fabrics impregnated with petrolatum wax, corrosion inhibitors, mineral fillers, and thermal extenders provide a further option for control of corrosion under pipe insulation. Two-part mastic coatings, most often aluminium-filled high solids epoxies, are used frequently on aged coating surfaces and on rusted steel that can only be hand or power tool cleaned. Electrochemical protection through the use of aluminium as sacrificial anode can also be provided by aluminium foil wrapping which in the case of austentitic or duplex stainless steel pipework prevents external SCC as well as pitting. Alternatively, thermal spray aluminium provides galvanic protection for both carbon and stainless steel, although the inherent porosity means epoxy, EPN or aluminium filled silicon sealer coatings are needed where immersion is likely.

Where corrosion is heavy, and it is not practical to replace badly damaged steelwork quickly, if at all, repair and reinforcement with the metal repair pastes considered in Section 3.7 either with fabric tapes, steel doubler plates or metal jackets and clamps provides strengthening for damaged areas as well as protection from further deterioration. Figure 3.22 provides a qualitative assessment of the heat resistance of the various anti-corrosive coating systems applied to protect steel onto which thermal insulation or protective fireproofing is applied.

Where equipment is failing and cannot be taken offline, in-service rebuilding can be made to manually prepared hot steel with special one-part epoxy repair pastes with reinforcement tape, plates or jackets. Related one-part (1K) epoxy and diverse two-part (2K) epoxy,

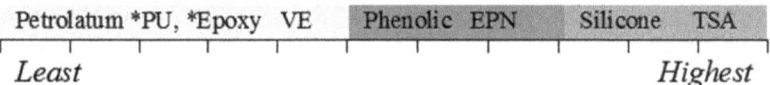

Figure 3.22   Heat resistance of anti-corrosive coatings for CUI and CUF *reduced when formulated as mastic coatings or extended with hydrocarbons.

Figure 3.23 Hot surface tolerance for in-service application of anti-corrosive repair materials for CUI and CUF.

epoxy-phenolic, modified silicone and proprietary inorganic anti-corrosive coatings are also available for application directly to hot steel to avoid full process shutdowns during planned maintenance. Figure 3.23 provides a qualitative assessment of the hot surface tolerance for in-service rebuilding/repair paste options.

There are also innovative solutions reported which combine corrosion resistance and insulation in a single-coating application in a major departure from the usual multi-step approach to prevention of CUI ((i.) conventional coating system on metal to prevent corrosion; (ii.) thermal insulation to reduce heat transfer and act as a vapour barrier; (iii.) cladding to protect insulation from damage and water ingress). It is beyond the scope of this book to consider the many different types of insulation materials or the traditional cladding/jacketing techniques used to keep them in place; it is pertinent, however, to review polymeric encapsulation techniques which seal and protect both new and existing lagging materials.

A zero permeability absolute vapour barrier solution to insulation cladding and jacketing of cylinders or spheres involves application of multi-ply laminates of aluminium foil and flexible thermoplastic polymer film, with and without fabric reinforcement, which are held in place with pressure-sensitive acrylic adhesive. These are, however, only effective on simple geometries without protrusions, and only if the edges are sealed carefully with zero permeance pressure sensitive adhesive tape. The protection of traditional lagging/insulation tailored to fit pipe spools or vessels and equipment with taps, inlet nozzles and supporting brackets is most effectively accomplished with thermoplastic polymeric resin dispersions in solvents or water which dry to form elastomeric water impermeable and seamless encapsulating membranes. Solvent-based synthetic rubber mastics and chlorinated acrylic copolymer emulsions which are non-porous when dry are used on equipment operating at cryogenic temperatures where it is critical to prevent water vapour ingress from the atmosphere through the insulation to cold steel. Waterborne acrylic, styrene acrylic, and proprietary chlorinated addition copolymer emulsions are used where microporosity is required to enable trapped moisture to escape from underneath or within the insulation to the atmosphere.

3.8.2   Fit-for-purpose Testing of Anticorrosive Coatings

NACE Standard SP0198 classifies protective coating systems for carbon and stainless steel under thermal insulation and fireproofing based on temperature capabilities determined from evaluation of the permeability and degradation resistance in cyclic heating and immersion exposures. Here the ISO 4628 corrosion, blistering, adhesion loss and ISO 16773 coating impedance loss is determined following heat exposure, cooling in air or quenching in cold water, then immersion in hot electrolyte or corrosive environmental exposures such as salt spray. The ASTM G189 Standard describes laboratory apparatus and procedures to simulate CUI exposure on coated hot-wall surfaces underneath thermal insulation, at either isothermal or cyclic temperatures, and with controlled delivery of ionic solutions to produce wet or wet-dry conditions. Corrosion rates with this CUI cell technique are generally higher than conventional panel immersion tests conducted in open or closed systems, so the relative effectiveness of protective coatings and influence of different insulation materials can be assessed in a short period of time.

Recommended Reading

- *BS 5970 Code of Practice for Thermal Insulation of Pipework and Equipment in the Temperature Range of −100 °C to +870 °C*, British Standards Institution, 2001, ISBN 0 580 33318 3.
- L. J. Korb, *Metals Handbook, Vol.13 Corrosion,* ASM International, 9th edn, 1987, p. 13, ISBN-10: 0871700190; ISBN-13: 978-0871700193.
- M. J. Mitchell, *Corrosion Under Insulation – New Approaches to Coating and Insulation Materials,* NACE International Corrosion, San Diego California, 2003, pp. 16–20.

3.9   Fire Protection

This section deals with passive fire protection (PFP) coatings for load bearing steelwork, and also fire resistant coatings for weatherproofing insulation used as part of systems to protect steelwork from corrosion.

One of the many advantages of steel is that it is non-combustible, but unfortunately above 550 °C (1000 °F) in a fire situation it begins to lose strength and can collapse under load. As such, fireproofing protection is mandatory for load bearing steelwork in buildings and structures where stability needs to be maintained for a reasonable period in the event of a fire to enable occupants and workers to escape and emergency services to respond. There are many different approaches to PFP/fire resistive coating protection for structural steel, they include: dense concrete encasement or filling; spray applied light weight cementitious, plaster and mineral fibres or boarding made from calcium silicate, gypsum and mineral fibre; and the main interest of this section, polymeric intumescent coatings. The selection depends upon many factors, not least construction design and fire risk nature – commercial or cellulosic fires fuelled by combustibles such as wood, paper, textiles reach 500 °C [932 °F] within 5 minutes and rise to in excess of 1100 °C [2012 °F] over time; petroleum or hydrocarbon fires fuelled by oil and gas with a very rapid heat rise to 1000 °C [1832 °F] within 5 minutes, and on to 1100 °C [2012 °F] quickly thereafter. Under the influence of fire, all of the above fire protection systems provide or form a heat insulation layer to delay the point at which underlying steel reaches its critical temperature.

In contrast, fire retardant polymeric encapsulation systems used to weatherproof insulation applied over steelwork in highly corrosive industrial environments are required to limit the flame spread, smoke development and fuel contributed that would otherwise occur with combustible polymeric coatings under the influence of fire.

Fact File: Flammability of Organic Materials

Organic materials burn in a multi-stage process and cycle involving:

- Heat from external sources;
- Decomposition producing highly reactive $^\bullet$OH and $^\bullet$H radicals which are responsible for flame spread in the combustion process;
- Degradation leading to the formation of flammable low molecular weight polymer fragment gases; which
- Diffuse to the surface where they mix with air and undergo ignition by nearby flame or by auto-ignition, leading to combustion; and
- Release of heat which propagates the cycle.

Reducing flammability of organic polymers can be achieved by inclusion of materials which:

- Chemically interfere in the mechanism of the chain reaction of the radicals in the oxidation and burning process;
- Produce large volumes of incombustible gases to dilute the oxygen content of the atmosphere surrounding the combustible material;

- React or decompose or change state endothermically absorbing heat to cool down the polymer;
- Develop an impervious fire resistant char on the surface to limit heat transfer and oxygen access.

3.9.1 Essential Chemistry and Technology

Passive fire protection contains ingredients that remain inactive until a fire occurs. Concrete and mineral boarding building materials along with spray applied lightweight mineral overlays insulate against heat transfer and are also fundamentally endothermic in nature as on exposure to high temperatures they release chemically bound water to cool the unexposed side and retard heat transfer over and above the inherent heat insulation effect of the original material. Once all the chemically bound water is released, the temperature on the un-exposed side tends to rise rapidly. Particulate mineral ingredients, which are also commonly included, undergo significant expansion when heated decreasing the density and providing additional heat insulation.

In contrast, intumescent coatings are designed to develop insu-lation properties and char on exposure to temperatures below which steel is adversely affected. Insulation is created as a result of a swelling in coating volume and subsequent decrease in density as well as by the formation of a layer of low-conductivity char on the exposed surface which also contributes to the retardation of heat transfer to the unexposed side. There is an ever increasing diversity of chem-istries and technologies employed to formulate intumescent coatings, although thin film intumescent coating systems favoured for fire-proofing steel are paint like materials which generally comprise a conventional primer, an intumescent basecoat, and a conventional finish coat—primers as described in Section 3.2 ensure a good bond is formed and provide first line corrosion protection for the under-lying steel; finish coats as discussed in Section 3.3 provide the re-quired surface appearance and day-to-day environmental resistance of the system. Intumescent basecoat compositions essentially contain an organic film-forming resin binder, an additional carbon source and catalyst which combine to form a carbonaceous soft char, and a blowing agent (spumific) to promote foaming and expansion of the carbonaceous char, and in some cases organic or inorganic particu-late materials which expand significantly on exposure to heat.

Film-forming resin binder selection for intumescent coatings ran-ges from water-based and solvent-based single part thermoplastic acrylics, to solvent-based and solvent-free two part epoxies amongst a wide variety of proprietary alternatives. Water-based intumescent coatings are less tolerant of humidity and low temperatures, and solvent-based systems are normally reserved for exterior on-site applications. Solvent-free, two-part epoxy systems are favoured for their versatility in use for interior and exterior site applications, where the highest level of mechanical properties and corrosion performance is required, and where steel requires protection from cellulosic fire or hydrocarbon fire. It is interesting to note that thin film intumescent coatings were originally limited to cellulosic fire protection and thick film intumescent coatings to hydrocarbon fire protection, although developments have enabled cross-overs to be made.

Fact File: Intumescent Coatings

Soft char intumescent coating systems on heat exposure at temperatures up to ~ 250 °C/~ 500 °F produce a light char which is poor conductor of heat and so retards heat transfer through the coating to the structural steel. Mineral hydrates are commonly incorporated as they release water vapour which doubles as a blowing agent as well as contributing endothermic cooling to the developing char. Fire resistant sealants based on silicone, silyl polyether and acrylic resins, along with fire rated polyurethane foams amongst other firestop technologies which are used as part of passive fire protection systems to seal joints and openings, also rely on soft char formation but function primarily by swelling when exposed at somewhat lower temperatures (~ 125 °C/~ 250 °F) to prevent the passage of fire and smoke in the early stages.

   Hard char intumescent coatings required for plastic pipe protection are also very poor conductors of heat and are created by the inclusion of graphite flake or sodium silicates. These ingredients not only expand in volume significantly on exposure to heat, they do so with considerable forces which are sufficient to squeeze together and shut melting/burning plastic pipes over which they have been applied.

Non-intumescent fire retardant coatings for the protection of insulation applied to steel operating equipment and piping at both high and cryogenic temperatures are typically made with thermoplastic polymeric resin dispersions in water or solvents, which dry to form elastomeric water impermeable protective membranes. These can be either microporous so as to enable trapped moisture to escape from underneath or within the insulation as vapour, or non-porous to prevent water vapour flow from the atmosphere through the insulation to cold steel. Both approaches lend themselves to the inclusion of ingredients which limit the flame spread and smoke development

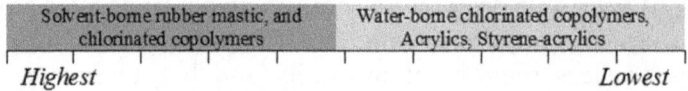

Figure 3.24 Environmental impact of installation of fire protection polymeric resin types.

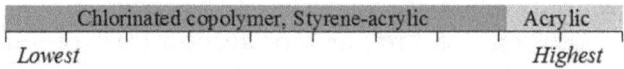

Figure 3.25 Environmental degradation resistance of fire protection polymeric resin types.

as well as fuel contributed to decrease the potential fire load and flammability. As identified in Section 3.8, acrylic, styrene-acrylic, and proprietary chlorinated addition copolymer emulsions are suitable where microporosity is required; solvent based synthetic rubber mastics and other proprietary chlorinated copolymer emulsions are needed where a total vapour barrier is needed. In terms of resistance to oxidation and weathering, acrylics provide the better overall performance, but in practice waterborne styrene acrylics are preferred for their speed of development of waterproofing properties and non-hazardous nature. The environmental impact of the various options from an installation perspective is indicated in Figure 3.24, and a qualitative assessment of the resistance to degradation on exposure to sunlight, water and temperature of fire protection polymeric resin types presented in Figure 3.25.

Fact File: Fire Retardant Resins and Additives

Fire retardant polymeric coating systems based on chlorinated resins, or systems including halogenated additives in combination with synergists such as antimony trioxide, on heating evolve HCl or HBr which react with the highly reactive •OH and •H radicals to form less reactive •Cl and •Br radicals and thereby contribute to flame quenching. This suppression of fire or slowing of the oxidation process does however result in the production of soot in smoke from incompletely burned material unless further additives such as phosphate esters which favour char formation, or highly surface active oxides which promote the oxidation process are included. Complete combustion of organic material is desirable as it leads to the production of simple molecules such as CO_2, H_2O, H_2, N_2 and SO_2 rather than soot, and also because incomplete combustion can lead to the formation of toxic HBr and HCl and highly poisonous CO and HCN amongst other more complex and noxious materials.

3.9.2 Fit-for-purpose Testing

Fire retardant coatings are required to stop flame and fire spreading over protected surfaces and are tested against BS 476 Parts 6 and 7, ASTM E84 and E2058 standards. BS 476 Part 6 and ASTM E2058 measure fire propagation index or characteristics; BS 476 Part 7 and ASTM E84 measure surface flame spread index and smoke developed index.

A Class 1 material provides the greatest resistance to surface spread of flame from the BS 476 Part 7 test which classifies materials as 1, 2, 3 or 4—ASTM E84 classifies A, B or C with a Class A having the lowest flame spread. A Class 0 fire rating for a coating material requires a Class 1 surface spread of flame classification (BS476 Part 7) and a ''very low'' fire propagation index (BS 476 Part 6).

Intumescent coatings, which are required to develop an insulating char to protect structural steel from heat generated by commercial or cellulosic fires are tested against BS 476 Parts 20, 21 and 22, BS EN 13381-8 and ASTM E119 amongst other national standards. These provide fire resistance duration ratings from 30 minutes to 180 minutes, with the most common periods being 60 minutes, 90 minutes or 120 minutes—a 90-minute fire retardant coating holding a fire back for 90 minutes before it breaches through the protective surface coating. Intumescent coatings designed to provide protection from petroleum or hydrocarbon fires are also evaluated against ANSI/UL 1709 for resilience in rapid temperature rise fire tests, and against ISO 22899 for resistance to jet fires.

Recommended Reading

- *Fire Protection for Structural Steel in Buildings (The Yellow Book)*, Association for Specialist Fire Protection, UK, 5th edn, 2010, ISBN: 9781870409254.
- *ETAG 018 Guideline For European Technical Approval Of Fire Protective Products Part 1*, European Organisation for Technical Approvals, Brussels, 2013.
- *RP 2218 Fireproofing Practices in Petroleum & Petrochemical Processing Plants*, American Petroleum Institute, Washington DC, 3rd edn, 2013.

References

1. B. A. Averill and P. Eldredge, *Electrochemistry, General Chemistry: Principles, Patterns and Applications*, Prentice Hall, 1st edn, 2007.

2. L. G. S. Gray, B. Drader, M. O'Donoghue, R. Garrett, R. Graham and V. J. Dratta, Using EIS to Better Understand Tank Lining Performance in Laboratory and Field Evaluation, *Corrosion*, 2003, 03382.
3. J. B. Enns and J. K. Gillham, *J. Appl. Polym. Sci.*, 1983, **28**, 2567–2591.

4 Rubber and Plastic Bonding, Repair and Restoration

4.1 Introduction

Many naturally sourced and derived materials have been used, and even more polymeric materials have been synthesised to satisfy specific requirements for industrial applications not met by concrete, masonry, wood and metallic engineering materials (covered in preceding chapters), nor for glass and ceramics (covered in the final chapter of this book). High molecular mass synthetic and semi-synthetic polymers, generally referred to as plastics, range from natural and synthetic rubbers, to thermoplastic or thermoset polymers and polymer matrix composites. Most engineering-grade rubbers and plastics require securing into place with a polymeric adhesive, and many suffer wear and damage in service, which can be restored with polymer-based solutions, and the following sections describe possible bonding and repair applications.

Fact File: Elastomers, Thermosets and Thermoplastics

Plastics are categorised typically as elastomeric, thermoset, or thermoplastic polymers:

- Elastomeric/rubber polymers, or elastomers, as they are also generally known, have randomly distributed branched crosslinked hard domains and flexible soft domains of coiled linear chains held together by Van der Waals intermolecular forces. The coiled chains are capable of unravelling under load in an elastic deformation and will return to

Industrial Polymer Applications: Essential Chemistry and Technology
By William R. Ashcroft
© William R. Ashcroft 2017
Published by the Royal Society of Chemistry, www.rsc.org

their original shape after release of a non-destructive load; heating above their T_g leads to degradation of the crosslinked domains within the cured matrix;

- Thermosetting polymers, or thermoset plastics, form 3D crosslinked networks that retain their shape when heated, as once "set" molecular flow is precluded and breakdown occurs if heated above their temperature limits—they cannot be reshaped above their T_g because they degrade before they melt;
- Thermoplastic polymers, or thermoplastics, have linear and branched chains which are not normally three-dimensionally crosslinked, so are therefore able to move past and over one another when heated above the T_g when Van der Waals attractive forces between the chains break down. When cooled, the melt can be reformed into a new shape, which means that in principle thermoplastic polymers can be melted again and recycled to form new items, but thermal processes inevitably cause some chemical degradation, especially with the higher melting engineering and high-performance plastics, and cooling can also lead to changes in alignment and folding of the polymer chains and therefore their structure and physical properties.

The structure of thermoplastics can be categorised either as:

- Amorphous—where long polymeric molecular chain segments are in disordered or irregular, folded and entangled coils, which are typically hard and brittle below their T_g. As temperatures increase close to the T_g, the molecular segments can begin to move, and above the T_g the mobility is sufficient to permit the polymer to flow as a highly viscous liquid; or
- Semi-crystalline—with regions of aligned (crystalline) polymer chains separated by amorphous folded regions. Strong intermolecular forces in the crystalline domains tend to prevent softening even above the T_g, so higher degrees of crystallinity result in polymers with higher stiffness and melt temperatures.

3D-printing technology involving fusion deposition of high-performance thermoplastics to make production-grade engineering prototypes and components is currently evolving rapidly and is increasingly likely to displace many of the more traditional metal repair, replacement and protection technologies discussed in Chapter 3 of this book! The thermoplastics in use for fusion deposition—ABS (acrylonitrile-butadiene-styrene), PC (polycarbonate) and PEI (polyetherimide)—with and without fillers to strengthen, bulk or add other properties, are progressively tougher and more durable, chemical, heat and flame resistant, so are increasingly likely to be adopted for making replacement parts for metals.

There are also developments in thermoset plastics technology and processes, whereby reactive liquid resins mixed with hardeners and catalysts on controlled and contained heating form crosslinked permanent organic networks which can, if required, be repeatedly reshaped like silica glass—this is due to reversible covalent bond exchange reactions that take place on reheating above the T_g. The technology promises to lend itself not only to self-healing repair to restore initial mechanical integrity, but also to recycling by means of grinding down into a powder and remoulding or injection to create new components and possibly even coatings, according to the originators.

Recommended Reading

- *CRC Handbook of Chemistry and Physics* ('Rubber Bible' or 'Rubber Book'), ed. W. M. Haynes, CRC Press, USA, 95th edn, 2014, ISBN: 1-4822-0867-9.
- *Natural Rubber*, ed. J. van der Heijden, R-S Information Center for Natural Rubber, Delft, The Netherlands, Newsletter 13, 1999, pp. 1–18.
- *Composites Engineering Handbook (Materials Engineering)*, ed. P. K. Malik, CRC Press, USA, 1st edn, 1997, ISBN-10: 0824793048, ISBN-13: 978-0824793043.
- D. Montarnal, M Capelot, F Tournilhac, L. Leibler, *Science*, 2011, **334**, 965–968.

4.2 Rubber Bonding, Repair, Rebuilding and Resurfacing

The unique performance and combination of properties of cured natural rubber has been exploited in industrial and engineering applications since the end of the 19th century, when the energy absorption and damping characteristics were first identified. Rubber derived from natural latex also offers good elasticity and resistance to moisture and is particularly versatile, as it can be compounded and vulcanised to give a bespoke range of hardness and stiffness. Synthetic rubbers made from petroleum-derived monomers have comparable mechanical properties to natural rubbers, but with better resistance to oils, chemicals, temperature and UV light. This section describes the various applications for polymers when natural or synthetic rubber engineering materials are fixed/sealed into place, as well as treatments for repair, replacement and restoration of damaged or worn rubber.

Fact File: Natural and Synthetic Rubber

Natural rubber obtained from the milky latex secretion of the tree *Hevea brasiliensis* is a natural polymer made from isoprene monomer [$CH_2=C(CH_3)-CH=CH_2$] linked in a *cis*-1,4-arrangement in loosely twisted chains, which gives rubber its highly elastic character. In contrast, gutta-percha, which is another natural polymer made from isoprene monomers has its links in the *trans*-1,4-configuration that makes it a crystalline solid at room temperature. Natural rubber is both elastomeric (soft and springy, deformable and will return to its original shape after deformation forces removed) and thermoplastic (weak forces between polymer chains easily broken by heating so polymer will flow/melt/unfold and can be moulded into a new shape). Natural rubber has a molecular weight of 100 000 to 1 000 000 Da so the long polymer chains must be partially broken down to achieve a viscosity low enough to permit compounding with fillers, pigments, antioxidants, plasticisers, accelerators and crosslinking agents prior to milling, then extrusion or moulding before being vulcanised (heat cured). Once rubber is vulcanized, it becomes a thermoset (strong covalent bonds between polymer chains that cannot be easily broken, which is why the polymer keeps its shape on heating), so rubbers are customarily vulcanised to achieve a balance of the properties of both thermoplastic and thermoset polymers, such that when heated they degrade rather than disintegrate.

Synthetic polyisoprene rubber can be made with properties very close to natural rubber by the solution polymerisation of pure isoprene monomer derived from petroleum sources in the presence of stereospecific catalysts to generate $\sim$99.5% of the *cis*-1,4-polyisoprene and minimal amounts of the *trans*-1,4 or the 1,2 and 3,4 addition isomers. Synthetic natural rubber along with *cis*–polybutadiene, whose properties are also close to that of natural rubber, are often blended with other synthetics, but most synthetic rubbers are made directly by copolymerisation to achieve specific properties. The first copolymers were styrene-butadiene rubbers (SBRs) with elastomeric properties close to those of natural rubber—these were supplemented by a number of other specialty elastomers, such as acrylonitrile-butadiene rubbers (also known as nitrile rubbers or NBRs), isobutylene-isoprene copolymers (butyl rubbers), and terpolymers of ethylene, propylene and diene monomers (EPDM rubbers). Neoprene, which is made by homopolymerisation of chloroprene [$CH_2=CCl-CH=CH_2$] is $\sim$90% *trans*-1,4-polychloroprene and is particularly useful at elevated temperatures and for heavy-duty applications. There are numerous other chemical types, which are classified and designated within the ISO 1629 system of nomenclature for synthetic rubbers. In addition to other wholly organic homopolymers, copolymers, terpolymers and thermoset PU elastomers, there are silicone rubbers and polysulfide rubbers, which have organic as well as inorganic contents. The properties available by design for rubber are seemingly limitless.

4.2.1 Essential Chemistry and Technology

Rubber bonding. Adhesives, sealants and gap fillers, which are capable of forming an adhesive bond to natural and synthetic rubbers, typically are also able to form strong bonds with many other engineering material types, such as metal and plastic. Solvent-based contact adhesives and sealants or gap fillers, made by dissolving solid natural or synthetic rubbers in a solvent, are able to form resilient bonds with most kinds of rubber—not only does the solvent content help penetrate and swell a rubber substrate surface, it also draws out any processing oils or waxes and absorbs them into the bonding rubber, which deposits on the contacting surface as the solvent evaporates. The bonding rubbers used may or may not be matched to the rubber substrate, and range from one-part natural rubber, neoprene and nitrile to two-part fluoroelastomers with proprietary curing technologies. Waterborne emulsion variants, with or without added solvent also solidify by drying, but are not suitable generally for sustained loading.

Solvent-free rubber bonding systems require non-porous surfaces to be manually prepared or chemically etched for optimum adhesion—solvents are still required, however, to degrease a rubber surface before and after surface roughening, but only if they do not draw processing oils and waxes to the surface, as this would disrupt the intended adhesive bond. One-part cyanoacrylate adhesives, which cure through reaction with moisture on the surface, are particularly suited to bonding close-fitting rubber joints, where very high bond strengths develop. Two-part PU adhesives provide strong flexible bonds between rubber and many other different surfaces and are particularly effective where vibration is likely to occur. Two-part epoxy adhesives, toughened with rubber-like micro-particles dispersed throughout the polymer matrix, provide outstanding adhesion to many rubbers and metals and are used where high peel strengths are required. Two-part epoxy/PU hybrid adhesives form strong and highly flexible bonds to many rubbers and metals with robust resistance to thermal cycling and shock.

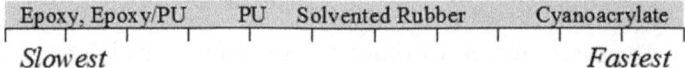

Figure 4.1 Speed of cure of rubber bonding adhesive technologies.

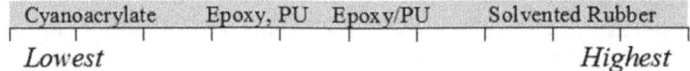

Figure 4.2 Peel strength of rubber bonding adhesives technologies.

Figure 4.1 presents a qualitative assessment of the relative speed of cure, and Figure 4.2 provides a qualitative assessment of the relative adhesive peel strengths of rubber bonding adhesives.

Rubber repair, rebuilding and resurfacing. Hardness, along with performance under tension and compression, are the primary measures used to define the practical limitations for thermoset and thermoplastic rubbers, so when it comes to the repair of holes, tears, splits, rips, gouges, wear or abrasion damage, then rebuilding materials are required with corresponding mechanical properties to the original rubber. Two-part PUs and PU/polyurea hybrids are used widely for cold vulcanising repair, rebuilding and even casting of replacements, as they can be formulated to meet rheological requirements for application and match the variety of hardness, tensile and compression properties found in most rubbers selected for industrial engineering applications. They are quick-setting and permit fast return to service even when used with one-part solvent-based moisture-cure PU primers, which are needed to ensure effective bonds. Two-part PU/epoxy hybrids can also be formulated for repair and rebuilding of rubber and, although they combine the flexibility and resilience of an elastomeric PU with the strength and adhesion of an epoxy, adhesion-promoting primers are still required for optimum performance.

Where resurfacing of rubber components is required to provide long-term or sacrificial protection from abrasion, corrosion, erosion and cavitation in localised pressure areas, coating-grade two-part PUs, PU/polyurea hybrids, polyureas, and PU/epoxy hybrids with the ability to dissipate heat and energy associated with friction, wear and cavitation forces are used, along with special water-resistant primers for immersed applications.

Figure 4.3 presents a qualitative assessment of the high temperature capability/heat resistance, and Figure 4.4 provides a qualitative assessment of the low temperature resilience of rubber restoration materials.

Figure 4.3 Heat resistance of rubber restoration technologies.

Figure 4.4 Low temperature resilience of rubber restoration technologies.

4.2.2 Fit-for-purpose Testing

Where rubber is being bonded to a rigid metal or plastic substrate, adhesion performance under tension, peel and shear can be verified from: the ASTM D429 Method A tensile strength test for a rubber part assembled between two parallel metal plates; the ASTM D429 Method B 90° peel stripping test for a rubber part assembled to one metal plate; and the ISO 1827/ASTM D429 Method H test for rubber-to-metal bonded quadruple shear components. In addition to providing upper limits for adhesion under tensile, peel or shear stress, the type or mode of rupture also helps confirm whether the strength of the polymeric bonding agent is higher than that of the cohesive strength of the rubber involved. Where rubber is being bonded to another rubber, adhesion performance under peel and shear can be verified from: the ASTM D413 standard 180° peel test between plies of fabric embedded into each of the rubber specimens bonded together; and the ASTM D1002 method, with single lap shear bonds between rubber samples, which are themselves bonded to metal at the lap joint to provide rigid support and stress distribution over the bond area.

Vulcanised and thermoplastic rubbers are categorised by their international rubber hardness degrees (IRHD) or Shore A durometer apparent hardness, so polymeric repair, rebuilding and resurfacing materials need to have their hardness evaluated in tests defined in the ISO 48/ASTM D1415 (IRHD) and ISO 7619/ASTM D2240 (Shore) standard methods. The apparent hardness of rubber bonded on to metal rollers is determined in the corresponding ISO 7267-1 (IRHD) and ISO 7267-2 (Shore) methods for curved surface test pieces.

Performance under tension for polymeric repair, rebuilding and resurfacing materials and rubbers is verified from the maximum tensile strength reached in stretching a test piece, the ultimate elongation or stretch at the moment of break, and the stress required to produce a given elongation as determined in the ISO 37/ASTM D412 test standard methods. Young's modulus (elastic modulus), which is the susceptibility to deform, is determined from the ratio of tensile stress to strain according to the ASTM E111 method for materials that exhibit both linear and nonlinear elastic stress-strain behaviour.

When a constant load is placed on an elastomer, compression occurs and the deformation is not constant but increases gradually over time, causing the material to relax or creep; it can even lead to physical and/or chemical changes which may prevent the material returning to its original dimensions after release of compressive stresses. The extent of the deformation or change depends not only on the time of the compression and recovery period but also the temperature, so the ability of a rubber to recover its original form, known as compression set, is determined in line with the ISO 815/ ASTM D395 (ambient and elevated temperatures) and ASTM D1229 (low temperatures) standards. These test methods are used to confirm polymeric repair, rebuilding and resurfacing materials behave in the same way as the rubbers being repaired or replaced. Conduct under compression for elastomeric rubber compounds other than those classified as hard rubber or sponge rubber is covered by ASTM D575.

Other crucial test methods used to verify fitness-for-purpose of rubber repair, rebuilding and resurfacing materials include: tear strength, which is a measure of the ease with which a tear can start and propagate, determined by the ASTM D624 test standard method; wet and dry abrasion resistance determined by a variety of methods covered in ISO 4649 (rotating drum), ISO 5470 (Martindale), ASTM D2228 (Pico), ASTM D4060 (Taber) standards; and rebound resistance determined by ASTM D1054, ASTM D2632, or ISO4662. Important test methods which confirm the electrical behaviour are: volume resistivity determined by the ASTM D991 test method to predict or confirm conductive or antistatic behaviour; dielectric strength determined by the ASTM D149/IEC 60243 method as a measure of the electrical strength as an insulator; and, dielectric constant (permittivity) determined by the ASTM D150/IEC 60250 method as a measure of the ability of an insulator to store electrical energy.

It is not only important for repair, rebuilding and resurfacing materials to have corresponding physical, mechanical and electrical properties to the original rubber, it is vital that the properties are retained or change in sympathy during operational use. Consequently, property retention testing after chemical, thermal (elevated and cryogenic) or other external pressure exposure, including cycling where appropriate, is an integral part of fitness-for-purpose validation.

Recommended Reading

- B. G. Crowther, *Handbook of Rubber Bonding*, ed. B. G. Crowther, Rapra Technology Limited Publishing, UK, 2001, ISBN: 1-85957-394-0.

4.3 Plastic Bonding and Repair

Plastics are chosen extensively and increasingly by designers and engineers to replace metals and ceramics because they offer unique combinations of properties not available in those classes of materials. There have been significant advances in chemistry and technology since the earliest chemical modifications of natural rubber and the first assemblies of fully synthetic thermoset polymers made with monomers derived from fossil fuels. In addition to the standard grades of thermoplastic and thermosetting polymers which are limited typically by temperature to use below $100\,°C/\sim 212\,°F$, there is now a wide range of engineering-grade thermoplastic and thermosetting polymers with long-term service capabilities up to $150\,°C/\sim 300\,°F$, plus a progressive range of high-performance thermoplastics and thermosets with long-term service capabilities above $150\,°C/\sim 300\,°F$, rising in exceptional cases to $260\,°C/\sim 500\,°F$.

Engineering and high-performance plastics are combined frequently with modifiers, stabilisers, pigments, metal fillers, mineral fillers, fibre reinforcements, flame retardants or lubricants to meet specific application requirements, and this chapter covers the basic options provided by conventional cold-applied polymeric technologies in bonding and repair of compounded plastics and reinforced plastic composites.

4.3.1 Essential Chemistry and Technology

Bonding. Where structural bonding is an acceptable joining technique for engineering and high-performance plastic and composite components, it usually necessary to pre-treat the surface of the plastic in order to enhance surface activity. When bonding identical or similar thermoplastics, plastic welding by a variety of techniques which deliver heat to the bonding surfaces, or solvent cementing involving application of a suitable solvent to the bonding surfaces prior to clamping together can be a more effective method of joining than adhesive bonding providing the correct quantity of heat/solvent in the latter approach is used. For adhesive bonding of dissimilar plastics, or plastics to metals, ceramics or other materials, application of a solvent to the plastic is also required to clean and degrease the material surface prior to increasing the size of the mechanical surface by abrasion, or by physical activation of the surface with a flame, plasma treatment, electrical (corona discharge) treatment, or chemical etching. Pre-treatment of the plastic helps bonding with a polymeric adhesive, but the overall strength of a bonded joint also depends on the joint geometry/design, the cohesion of the bonding adhesive, and suitable surface preparation for any non-plastic adherends.

The choice of polymeric adhesive depends on the intended operating conditions as well as the materials being bonded, and many of the materials described in Section 3.7 for bonding metals are equally suitable for plastics and composites. Hot-melts are again useful for low stress product assembly and one-part cyanoacrylates for bonding small areas and closely fitting surfaces. Solvent-based rubber contact adhesives, toughened acrylics, and two-part PUs are selections where vibration is likely to occur. Two-part MMAs are widely used as structural adhesives in automotive and marine applications for their high strength and resistance to impact across a wide temperature range, and two-part epoxies are used for their retention of adhesion, strength and load bearing at elevated temperatures—heat-curing epoxies here providing higher temperature capabilities than traditional cold-curing epoxies. Figure 4.5 presents a qualitative assessment of the relative speed of cure of plastic bonding adhesive technologies.

Repair. Engineering and high-performance plastics and polymer matrix composites are susceptible to fatigue arising from cyclic or static stresses, and fracture from applied forces or impacts, which cause tiny cracks or other defects that are liable to spread rapidly and lead to sudden dramatic failures. Plastics and polymer matrix composites can also wear due to friction and abrasion, as well as age prematurely from over exposure to heat and (UV) light outside their design limit capabilities.

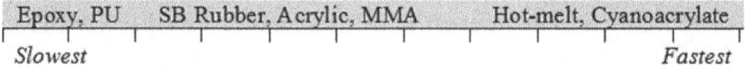

Figure 4.5 Speed of cure of plastic bonding adhesive technologies.

Surface cracks and defects in thermoplastics can in principle be remedied easily by re-melting and reforming the component so the polymer chains rearrange and intermolecular Van der Waals forces reinstate. The repair of wear and ageing, however, requires lost or degraded material to be replaced, and unfortunately conventional surface coating solutions intended to restore and protect from further loss are unlikely to survive for any useful period of time, unless they provide a monolithic structural solution. As such, the only effective alternatives to total replacement are to apply more of the same material, either by hot fusion welding or as a solvented cement filler which cold fuses into place to create a uniform plastic-on-plastic repair.

Fact File: Self-healing Polymers

The development of self-healing polymers that bond tiny microscopic cracks formed due to impact or fatigue established the possibility for "smart" coatings and plastics, which survive general, wear-and-tear. They have been followed by regenerating materials with "healing" agents embedded in parallel networks of capillary-like tubes which, when damaged, release liquids that combine to form a gel which fills large cracks and holes in both thermoplastics and thermoset plastics. This opens the possibility for future structural materials that are capable of continuously regenerating and remodeling in the same way human bone regenerates cells to ensure it remains strong and flexible, and forebodes of a future revolution in industrial engineering where there is no need for protection and repair materials!

Cracks, defects, wear and surface degradation of thermoset plastic parts or components generally require sealing or patch repair with the same kind, and preferably grade, of thermoset resin used to make the original component for optimum adhesion and, of course, to ensure monolithic cured behaviour. Two-part MMA, UPR, VE, PU and epoxies, with or without reinforcing fibres/fabrics/meshes and abrasive fillers can be used to provide long-term repairs for damage arising from upset conditions, or for sacrificial protection in continuously aggressive environments.

The repair of cracked, damaged, worn and degraded thermoplastic or thermoset polymer matrix composites can be quite complicated, as it depends on the extent of damage and nature of the constituents and structure – non destructive test (NDT) methods are usually undertaken in addition to visual inspection to identify splitting in laminate structures or puncture damage in sandwich structures, which can be more extensive than that seen from the surface.

Fundamentally, however, most repairs rely on adhesive bonding and the reduction of stresses in the damaged region to keep cracks from opening and growing.

The most basic type of repair can be made with two-part epoxy, UPR, VE or MMA coatings with non-structural fillers applied over the damaged surface until a more permanent repair can be made—cosmetic repairs of this type are only used though where composite strength is unimportant and there is an immediate need to protect from aggravating environmental factors. Injection with unfilled two-part epoxy, UPR, VE and MMA resins can also be effective in limited instances where delamination is restricted to a single ply, but not much strength is regained with this technique so it is, again, only used to slow the spread of delamination as a temporary measure.

Semi-structural repairs of sandwich punctures with thick, solid laminate skins which are able to take bolt loads are possible with mechanically fastened core plugs or bolts sealed in position with a non-structural patch repair coating over the damaged area. Here, two-part epoxy, UPR, VE and MMA patch repairs with non-structural fillers are used to protect both plug and core from exposure to anything in the environment which may cause further deterioration of mechanical performance.

Strength is critical for primary structural thermoset polymer composites that are operating close to their design limits so replacement of structural reinforcement and polymer matrix with either wet lay-up materials or a prepreg patch, using the vacuum bagging technique, is essential to ensure the required compaction and consolidation of repair plies. Lightly loaded thin laminate structures can be repaired with replacement fibre plies impregnated with liquid resins applied by hand roller or brush over and around the damaged area to form (after consolidation by air extraction, and heating if required to facilitate resin thinning and/or cure), a structurally bonded external doubler with a significant retention of the original strength of the composite, albeit with a stiffness and weight penalty. Neither the reinforcing fibres nor the resin need match the original composite material and two-part epoxy or phenolic resins are used predominantly since wet lay-up MMA, UPR and VE resins can lose too much of their volatile components during the vacuum extraction process for this method of repair.

In contrast, full structural strength repair of laminate structures requires replacement of the matrix resin and each ply of fibre lost from the area damaged and prepared for repair—this is best achieved with prepreg patches made from grades of resin and reinforcement comparable to the original composite. For a typical repair, a series of prepreg patches are cut to size to fill the repair area in a precise stepped 3D fit along with a final cosmetic layer which is slightly larger to allow for sanding down after vacuum bagging/curing to achieve a smooth surface finish. Prepreg patching either involves heat curing at temperatures high enough for thermoplastic resin matrices to melt together, or at temperatures high enough to make a thermoset resin matrix of the prepreg patch soften or briefly liquefy before bonding and crosslinking, or UV light curing of a photo-initiated thermoset resin matrix.

Fact File: Prepregs

Unidirectional tape, sheet and fabric reinforcements made from carbon, glass, aramid and other fibres, when pre-impregnated with a polymer resin matrix material are universally known as prepregs. Other than for repairs, they are mainly used with moulds or tools that are heated under pressure to produce primary composite components—prepregs are chosen for their ease of use and consistent properties, which are made possible by controlled mechanised manufacture, which ensures homogeneity.

Thermoplastic prepregs are typically made by coating assembled fibre reinforcements with thermoplastic resins that have been heated above their melting points and which, on cooling back to room temperature, can then be stored indefinitely without any change to physical form. Common resins used include PP, PET, PA, PEI, PPS and PEEK.

The more commonly encountered thermoset prepregs are made with a pre-catalysed resin matrix, which is used to impregnate and bond fibres together during a controlled partial cure reaction. Epoxies are frequently used as primary resin matrix as they part cure, or "B-stage", to a pliable solid state. Other thermoset resins made into prepregs include UPR, VE, PU, phenolic, cyanate ester, BMI and polyimide. B-Staged prepregs all continue to cure slowly at ambient temperatures so require refrigerated storage until needed when heat is applied to complete the crosslinking reaction.

The emergent technology for UV (A) light curing of photo-initiated thermoset UPR and VE resins lends itself well to glass reinforced laminates and prepregs which cure rapidly and completely, even at low ambient temperatures. They need to be stored away from sunlight and heat sources until ready to use.

4.3.2 Fit-for-purpose Testing

Where plastic is being bonded to another rigid plastic or a metal substrate, shear strength is verified using the ASTM D4501 block-shear method which minimises undesirable peel and through-thickness tensile stresses in the adhesive, and also by the ASTM D3164 lap-shear sandwich joint method. These are preferred over the ASTM D1002 single lap shear test where specimens made from plastics with lower tensile properties than metals can deform and fail during testing with high strength adhesives under load—the block-shear and lap-shear sandwich joint methods also provide reliable comparative shear strength data for different adhesives, surface pre-treatments and plastics. ASTM D3807 is the method used to determine peel strength properties of adhesives with engineering plastics, and ASTM D5401 is used to measure fracture strength in cleavage of bonded joints with reinforced plastic specimens.

Although there is no standard test protocol for evaluating the fatigue threshold of adhesively bonded plastic joints, Goodman diagrams or S–N curves which plot the cycles-to-failure (N) as a function of stress (S) required to cause fatigue failure under tension is the industrial and academic standard approach for engineering plastic and polymer matrix composite specimens. S–N curves are produced by testing groups of cured test specimens, firstly at a high peak stress where failure is expected in a fairly short number of cycles, then at decreasing test stresses until one or two specimens do not fail in the specified numbers of cycles, which is usually at least 10^7 cycles—the highest stress at which a non-failure occurs is taken as the fatigue threshold. Both "dog bone" specimens utilised in ISO527/ASTM D638 tensile strength, modulus and elongation testing of plastics, and straight-sided specimens required in ASTM D3039/ASTM D5083 tensile property testing of polymer matrix composites/reinforced thermoset plastics are used, with or without notches in fatigue testing.

In addition to ISO527/ASTM D638 tensile properties, the mechanical and physical fitness-for-purpose of unreinforced polymeric repair materials generally also requires determination of ISO 178/ASTM D790 flexural properties, ISO 604/ASTM D695 compressive properties, ISO 179/ASTM D6110 Charpy or ISO 180/ASTM D256 Izod impact resistance, ASTM D6671 fracture toughness, and ISO 75/ASTM D648 heat distortion temperature or ISO 11357-2 glass transition temperature. In contrast, fitness-for-purpose testing of reinforced wet lay-up or prepreg repairs is directed by the anisotropic nature of polymer matrix composites whereby physical properties such as stiffness may depend on the direction of applied force or load.

Some of the same tests used for unreinforced plastics already mentioned are relevant, although variations specific to polymer matrix composites are preferred like the ASTM D7264 flexural method which uses bars of rectangular cross section supported on beams with a greater span-to-thickness ratio than the ASTM D790/ISO 178 method for unreinforced plastics. ASTM D4762 is the standard which provides an extensive listing of all other methods relevant to the testing of laminate structure and sandwich structure composites which are beyond the scope of this book.

Recommended Reading

- *Applied Plastics Engineering Handbook – Processing and Materials,* ed. M. Kutz, William Andrew Inc., USA, 2011, ISBN: 978-1-4377-3514-7.
- M. J. Troughton, *Handbook of Plastics Joining: A Practical Guide,* William Andrew Inc., USA, 2nd edn, 2008, ISBN: 978-0-8155-1581-4.
- C. A. Harper and E. M. Petrie, *Plastics Materials and Processes: A Concise Encyclopaedia,* Wiley-Interscience, 2003, ISBN-10: 0471456039, ISBN-13: 978-0471456032.
- *Handbook of Thermoset Plastics,* ed. S. H. Goodman and H. Dodiuk-Kenig, William Andrew Inc., USA, 3rd edn, 2013, ISBN: 978-1-4557-3107-7.
- M. M. Abdel Wahab, *Fatigue in Adhesively Bonded Joints,* ISRN Materials Science, Hindawi Publishing Corp., 2012, 2012, Article ID 746308.
- R. P. Wool, Self-healing Materials: A Review, *Soft Matter,* 2008, **4**, 400–418.
- S. R. White *et al.,* Autonomic Healing of Polymer Composites, *Nature,* 2001, **409**, 794–797.

5 Glass and Ceramics Repair and Bonding

5.1 Introduction

Traditional ceramic materials, which include concrete and glass, along with advanced technical ceramics, are important industrial engineering materials that are used because they are: hard and less dense than most metals and alloys; more heat and corrosion resistant than metals and plastics; chemically and physically stable at high temperatures; and, not thermally or inherently electrically conductive. Some ceramic materials are more brittle than others, and more prone to thermal shock, therefore they may suffer damage in service which can, depending on circumstances, be restored with polymer-based solutions. Repair, restoration and protection solutions for traditional porous ceramic materials made from cement and clay-based masonry have been discussed already in Chapter 2; the following sections deal with technical and advanced ceramics, which are non-porous and not generally clay based.

Vitreous silica glass, probably the best known of all technical ceramics, has the same chemical composition as silica quartz, but the structural arrangement of the repeating blocks of SiO_4 differs. Silica glass, and all other materials exhibiting glass-like properties, are amorphous, with their backbone assembly randomly arranged, whereas silica quartz is crystalline, with a very ordered structure that is more mechanically and thermally stable than a disordered one. Ceramics can be made with fully crystalline ordered structures, or partly crystalline/partly amorphous structures, and many glasses can be transformed into ceramics by heating them above their glass

Industrial Polymer Applications: Essential Chemistry and Technology
By William R. Ashcroft
© William R. Ashcroft 2017
Published by the Royal Society of Chemistry, www.rsc.org

transition temperatures, shaping and then slow cooling to allow rearrangement from a random to a crystalline structure—so-called glass-ceramics made from materials originally manufactured as glasses have enhanced mechanical and thermal properties associated with the increase in structural order.

Advanced ceramics are made from metal oxides (typically alumina and zirconia) and non-oxides (borides, carbides, nitrides such as boron nitride, and silicides such as silicon carbide), with or without particulate or fibre reinforcement. Ceramic matrix composites with embedded whiskers (long multi-strand fibres) of crystalline silicon carbide or boron nitride, are typified by their enhanced durability and high mechanical strength at the highest continuous working temperatures, which are well beyond the capabilities of metal matrix composites and polymer matrix composites.

Recommended Reading

- C. B. Carter and M. G. Norton, *Ceramic Materials: Science and Engineering*, Springer Science & Business Media, 2nd edn, 2013, ISBN-10: 1461435226, ISBN-13: 978-1461435228.
- M. Bengisu, *Engineering Ceramics*, Springer-Verlag, Berlin Heidelberg, 2001, eBook ISBN 978-3-662-04350-9.
- K. K. Chawla, *Ceramic Matrix Composites*, Chapman & Hall, London, 1993, ISBN-10: 0412367408, ISBN-13: 9780412367403.

5.2   Fused Silicate Glass Repair

Silicate or silica glass, made by melting crystalline quartz sand at high temperatures, and fused as a thin amorphous layer on metal, ceramic or a glass, is known as vitreous enamel or porcelain enamel. The smooth, hard coatings of glass-fused-to-steel (GFS) vessels, pipes, valves, fittings and tanks have very low thermal expansion co-efficients, resist high temperatures and are inert—they are renowned for providing protection of underlying steel from corrosion from a wide range of oxidising and reducing chemicals. Although relatively scratch- and chip-resistant, silicate glass will crack or shatter if stressed, bent or mechanically impacted, but repair of localised damage is often possible and preferable to replacement of an entire glass lining.

Fact File: Glass Types and Their Uses

Borosilicate glasses, used in the manufacture of chemical glassware and corrosion-resistant pipe, drain and vent systems, are made from silica sand (SiO_2) and boron trioxide (B_2O_3) as the main glass-forming constituents with smaller quantities of soda (Na_2O) and alumina (Al_2O_3). (Sodium) borosilicate glasses are characterised by low coefficients of thermal expansion, lower vulnerability to cracking and greater dimensional stability than **soda-lime glasses,** which are made from silica sand, sodium oxide, alumina, magnesia (MgO) and lime (CaO), or **lead-oxide (crystal) glass** made from silica sand, lead oxide (PbO), potassium oxide (K_2O), soda, zinc oxide (ZnO) and alumina. Higher strength glasses drawn as fibres include **alumino-silicate glass** made from silica, alumina, lime, magnesia, barium oxide (BaO) and boron trioxide used in making glass-reinforced composites, and extremely clear **oxide glass** made from alumina and germanium oxide (GeO_2) for fibre-optic waveguides.

5.2.1 Essential Chemistry and Technology

Where cracks or holes extend through the glass lining to the steel, and where time permits and equipment can be moved into an oven, repair by regeneration to a thickness almost equal to the existing glass layer can be made successfully by the sol–gel process—repeated application of silicon alkoxide solution, hydrolysis and a thermal treatment to force polycondensation and form thin incremental film layers of interlocked replacement glass. Fusing molten quartz over the damaged area is a faster repair, albeit it requires significant accessibility and skill to accomplish.

Where hotwork is prohibited, and where complex geometries limit accessibility, effective repairs can be made by drilling and tapping studs into the steel around the repair area, applying a furan or silicate cement filler and a PTFE gasket, over which a tantalum or hastalloy patch is bolted into place. Where operating temperatures and the nature of the containment or process chemicals permits, priming of exposed steel and surrounding undamaged glass with a silane coupling agent and patching with polymerisable composite patches made from two-part epoxy, EPN or glycidylmethacrylate (GMA) resins can be used in place of rarer tantalum and exotic corrosion resistant alloy patches. Figure 5.1 provides a qualitative assessment of the resistance

Figure 5.1   Resistance to corrosion and heat of glass repair technologies.

to corrosion and heat for the various technologies used for the repair of fused glass.

Small cracks, chips and pinholes in glass-fused-to-steel can be filled with ceramic nano-particle filled cements and resins based on silicate, furan, GMA, epoxy or EPN resins—the selection of which depends on the equipment operating temperature and chemical exposures involved.

Larger defects which lead to through-wall corrosion to the steel exterior can often be repaired with the same ceramic filled cements applied internally to cold bonded external steel doubler plates or external composite patches or wraps as previously discussed in Section 3.7.

5.2.2 Fit-for-purpose Testing

Simultaneous validation of adhesion and resistance to specific chemicals at process temperatures for repairs made with filled cements, resins and composite patches can be effected through ISO 4624/ASTM D4541 pull-off strength tests on coatings or patches applied to glass-fused-to-steel panels or plates exposed to chemicals in NACE TMO174 atmospheric pressure and pressurized Atlas Cell tests. Otherwise, determination of the resistance to attack by acids, alkalies and neutral liquids at ambient, elevated or boiling temperatures is determined in accordance with the ISO 28706 test regime for vitreous and porcelain enamels.

Important physical property testing includes ISO 15695 scratch resistance and BS EN 15771 surface scratch hardness according to the Mohs scale, ASTM D4060 Taber/sliding abrasion resistance, and ISO 28721-3 thermal shock resistance. Determination of resistance to mechanical damage of repair materials is determined from the ASTM D256/ISO 180 notched Izod impact test.

Recommended Reading

- A. I. Andrews, S. Pagliuca and W. D. Faust, *Porcelain (Vitreous) Enamels,* Tipografia Commerciale srl, Italy, 3rd edn, 2011, ISBN: 978-88-903905-3-1.

5.3 Ceramic Bushing Repair and Restoration

Glass and porcelain ceramics are used widely in the manufacture of insulating bushing, posts and overhead line support components for electrical power transmission and distribution. Although glass has the highest dielectric strength, it attracts condensation, which can lead to flashover in contaminated locations, and it cannot be cast into bulky irregular shapes which are free from internal stresses. Fortunately, porcelain ceramics made by heating and moulding combinations of clay, quartz or alumina and feldspar do not suffer the same drawbacks and, when covered with a smooth glaze to aid the shedding of water, provide outstanding insulating properties, mechanical strength and impermeability, so have become the first choice for design engineers. There are other non-ceramic polymer composite materials used in the manufacture of insulation, and although they do not have a long-term proven service life, their higher mechanical strength to weight ratios and fracture resistance offer potential significant advantages compared to conventional ceramic insulation.

Porcelain insulation components tend to be bulky, heavy and brittle, so are frequently damaged accidentally in transit and handling or through intensive electrical discharges on polluted surfaces, or by environmental and mechanical stresses such as wind and ice, and even maliciously by vandalism. Rebuilding of broken components to make them safe by removing sharp edges, sealing the exposed surface against moisture ingress, and reinstating creepage distances, are all possible with polymeric repair.

5.3.1   Essential Chemistry and Technology

Porcelain ceramic repairs are easier to achieve than glass repairs, not just because exposed cracked porcelain has sufficient porosity to facilitate adhesion, but also from aesthetic considerations. Rebuilding chips and voids where pieces of porcelain are missing is possible with putties based on insulating two-part epoxy, PE, PU or silicone resins containing pigments to match brown or grey porcelain, and high filler contents which enhance dielectric strength. In contrast, repair of porcelain, or glass, where a broken piece is available, can be achieved with a two-part unfilled epoxy or one-part cyanoacrylate adhesive. Replacement of burned insulator glaze, or sealing with a coating after a putty rebuild, is achieved with a two-part PU varnish, a two-part silicone rubber, or a silicone grease coating, all of which act as preventative treatments against future surface contamination from pollutants.

5.3.2   Fit-for-purpose Testing

Validation of adhesion of bonding agents and repair putties to porcelain, and of non-grease coatings to the putties can be made from ISO 4624/ASTM D4541 pull-off tests.

The physical fitness-for-purpose of rebuilding materials requires determination of ASTMD2240/ISO 868 hardness, and heat resistance established from ASTM D3418/ISO 11357-2 glass transition temperature or ISO 75/ASTM D648 heat distortion temperature tests, so as to endorse thermal classification defined by IEC 60085 and NEMA/UL standards for electrical insulation. The toughness capabilities of rebuilding materials is verified from ISO 180/ASTM D256 Izod impact resistance, ISO 178/ASTM D790 flexural properties and ISO527/ASTM D638 tensile properties, especially E-modulus.

The most important testing to confirm the electrical behaviour of insulating bushing rebuilding materials and varnishes is ASTM D149/IEC 60243 dielectric strength—the voltage required to break down or arc materials. Other electrical fit-for-purpose tests for insulating materials are ASTM D150/IEC 60250 dielectric constant (the ability of an insulator to store electrical energy) and dissipation factor (the inefficiency of an insulating material to store electrical energy), as well as ASTM D257/IEC 60093 surface resistivity (the resistance to leakage of current along the surface of an insulating material) and volume resistivity (the resistance to leakage of current through the body of an insulating material).

Recommended Reading

- J. Liebermann, *High-Voltage Insulators: Basics and Trends for Producers, Users and Students*, Schulze KG, Lichtenfels, Germany, 2012, ISBN: 978-3-87735-210-6.
- M. T. Gençoğlu, *NWSA*, 2007, 2, 274–294 [The comparison of ceramic and non-ceramic insulators].
- P. Barber *et. al.*, *Materials*, 2009, **2**, 1697–1733 [Polymer composite and nanocomposite dielectric materials for pulse power energy storage].

5.4   Advanced Ceramics Bonding and Restoration

Advanced ceramics differ in composition to traditional ceramics and simple composite mixtures such as silicate glass, concrete and clay-based masonry/porcelain, by being compounded typically from metallic elements with carbon and nitrogen, so-called non-oxides, as well as with oxygen. The form and properties, like all materials, are dictated not only by the elements present, but also by the specific nature of the bonding between their atoms and the way they pack together.

The primary, and strongest, chemical bonds found in ceramics are a mixture of ionic and covalent types depending on the potential of an element's atoms to accept or donate electrons – where there are significant differences in electronegativity between elements, electrons are transferred between the atoms resulting in ionic bonds; where there are small differences in electronegativity between the elements, electrons are shared between the atoms in covalent bonds. There are also molecular-level van der Waals dipole–dipole attraction forces, which create secondary weaker bonds between adjacent atoms. Higher tier bonding occurring between closely packed positive metal ions gives rise to electrically conductive ceramics as there are free/delocalised electrons available—conventionally of course, ceramics have been designed with an absence of free electrons, so they are poor conductors of electricity and heat, so it seems there are effectively no limits to the properties with which ceramics can be designed.

The method of manufacture also has a significant impact on the microstructure of advanced ceramic materials – molecular alignment occurs during polymerisation and crosslinking of incremental layers in a sol–gel process, and through crystallisation by hydrothermal synthesis or by controlled cooling from a molten mass; otherwise, there is fusion at the atomic level, which is induced by compacting and forming during sintering under heat and/or pressure. It is evident that whenever atoms are allowed to achieve equilibrium separation, the energy state is a minimum, so most advanced ceramics are characteristically low density, high hardness, thermally stable and inert to chemical attack and corrosion. Toughening and increasing resistance to damage or failure from inherent brittleness is achieved by incorporating ceramic particles, whiskers or continuous ceramic fibres to form a composite where the advanced ceramic is the continuous matrix—these differ from cermet composites where the metal acts as binder or as metal matrix, and also differ from polymer-ceramic composites which incorporate ceramic fillers in an organic or organosilicon polymer matrix.

Advanced ceramics are increasingly important and versatile engineering materials, and there are a number of suitable polymeric solutions used with them for fixing components to other ceramics, metals and superalloys, or to other advanced composite materials. There is currently limited potential for the use of polymers for the repair of flaws and cracks that form on the surface or in the bulk of ceramics from friction, fatigue or overloading.

5.4.1 Essential Chemistry and Technology

Bonding ceramics to other ceramics can be achieved by co-firing, brazing or other proprietary fusion technologies, or with ceramic, graphite or silicate adhesives when long-term strength retention at temperatures over ~ 1000 °C/~ 2000 °F is required. Although thermally conductive cold vulcanised silicone adhesives have been used to bond ceramic tiles onto the space shuttle, conventional engineering grade thermoset and thermoplastic polymer adhesives are not suitable generally for prolonged use at temperatures above ~ 150 °C/~ 300 °F.

However, higher performance thermoplastics and thermosets with long-term service capabilities up to ~ 260 °C/~ 500 °F, which can survive short-term excursions to ~ 350 °C/~ 660 °F, are often used where brittle ceramic adhesives are not suitable and where operating temperatures permit. They are usually supplied as precast or pre-impregnated adhesive films, sheets and tapes based on B-staged acrylic, benzoxazine, bismaleimide (BMI), polyimide and polysiloxane resins, or as a liquid or paste made with BMI or cyanate ester resins.

Thermal healing to restore strength and resilience against operational stresses is unfortunately an inescapable requirement in the repair and restoration of advanced ceramics with flaws and damage cracks on the surface and in the bulk. Solid-state crack healing, involving heating with sufficient energy to regenerate solid contact between disrupted crack edges of the ceramic by fusion, does not need air or added repair materials. Certain ceramics with ternary MAX (M = early transition metal; A = A group element; X = C and/or N) elemental constitutions, including for instance ceramic matrices with embedded silicon carbide particles, are known to self-heal when operating at high temperatures in air through preferential oxidation of the A element which fills spaces between cracks and pores with an autoxidation product.

Although ceramic-filled polysiloxanes are used in the manufacture of injection moulded ceramic composites, there are currently no polymer-ceramic composite repair technologies known to penetrate, seal and reinstate cracks in damaged advanced ceramics.

5.4.2 Fit-for-purpose Testing

Where a ceramic is being bonded to another rigid ceramic, metal or alloy, adhesive strength can be verified using the ASTM D1002 single lap shear method, or the ASTM C1469 shear strength method, which is specific to joints between advanced ceramics. When co-bonded to thermoset or thermoplastic plastic components, the ASTM D3164 lap shear sandwich joint and the ASTM D4501 block-shear methods are again used to minimise peel and through-thickness tensile stresses in the adhesive. The bonding between ceramics using a core laminate adhesive is verified by ASTM C297 load transfer method which provides valuable comparative flatwise tensile strength data for different adhesives and surface pretreatment/preparation methods.

Important physical property testing for adhesives used to bond advanced ceramics starts with T_g melting temperature ranges measured from ISO 11357-2 differential scanning calorimetry or ISO 6721-11 dynamic mechanical analysis methods. Thermal oxidative behaviour is determined from ISO 11358/ASTM E1131 thermo gravi-metric analysis, and thermal expansion behaviour is determined from ISO 11359-2 thermo mechanical analysis or by the ASTM E831 method. ASTM F433 thermal conductivity measurements are taken to confirm heat transfer capabilities, along with ASTM E1004 electrical conductivity, and/or ASTM D149/ IEC 60243 dielectric strength and ASTM D257/IEC 60093 volume resistivity and surface resistivity measurements to verify electrical insulation properties.

ASTM D638/ISO 527-1 tensile strength/modulus/elongation, ASTM D790/ISO 178 flexural strength/modulus and ASTM D695/ISO 604 compressive strength/modulus of neat resins used in bonding are all mechanical properties which can be helpful to designers depending on the intended application. However, fracture toughness measured in the ASTM D5045/ISO 13586 edge notch bending and compact tension methods is the critical fit-for-purpose mechanical property for adhesive applications involving high static loads, cyclical loading and vibration.

Recommended Reading

- *Mechanical Properties and Performance of Engineering Ceramics II: Ceramic Engineering and Science Proceedings,* ed. R. Tandon, A. Wereszczak and E. Lara-Curzion, John Wiley e-book, vol. 27, issue 2, 2009.
- *Adv. Mater. Processes Magazine*, 2001, **159**(12), 155–162 [Guide to engineered materials].
- C. B. Carter and M. G. Norton, *Ceramic Materials: Science and Engineering,* Springer Press, 2nd edn, 2013, ISBN-13:978-1461435228, ISBN-10:1461435226.
- P. Griel, *J. Adv. Ceramics,* 2012, **1**(4), 249–267 [Generic principles of crack-healing ceramics].

Appendix 1   Glossary of Resins, Polymers and Plastics

Acrylic resins: acrylic acid (prop-2-enoic acid, $H_2C=CHCO_2H$) derivatives such as methyl methacrylate (methyl 2-methylpropenoate, $H_2C=C(CH_3)COOCH_3$) used with catalysts in the formulation of thermosetting acrylic and acrylate adhesives, flooring and coatings; typically toughened with acrylic polymers to enhance adhesion (see MMA resins below); free radical cure mechanism initiated by UV light, peroxide and amine, or other radical sources results in fast setting-times when compared with other resins systems.

    Acrylic plastics: thermoplastic polymers made from one or more derivative of acrylic acid; polymethylmethacrylate (PMMA, $(-CH_2-C(CH_3)(CO_2CH_3)-)_n$) is probably the best-known example, and is used for toughness, high transparency and resistance to UV and weathering; saturated acrylic polymer emulsions and solutions used in fast-drying film-forming coatings and adhesives.

    Acrylonitrile-butadiene-styrene (ABS) plastics: thermoplastic terpolymers made from acrylonitrile (2-propenenitrile, $H_2C=CH–CN$), buta-1,3-diene ($H_2C=CH–CH=CH_2$) and styrene (phenylethene, $C_6H_5CH=CH_2$); amorphous in nature with no true melting points and a glass transition temperature of approximately 105 °C/221 °F; used for their light weight, impact resistance and toughness in the production of injection moulded and extruded engineering components.

    Alkali silicate resins: solutions of sodium or potassium silicate glasses in water, also known as waterglass, with chemical formula $Na_2(SiO_2)_nO$ or $K_2(SiO_2)_nO$; characterised by low viscosity and ability to form highly crosslinked networks and tightly adhering inorganic bonds and insoluble silicate networks; used to penetrate and react

Industrial Polymer Applications: Essential Chemistry and Technology
By William R. Ashcroft
© William R. Ashcroft 2017
Published by the Royal Society of Chemistry, www.rsc.org

chemically within the pores and capillaries of concrete and masonry to block moisture penetration, or as a binder to lock metallic zinc powder into place in zinc-rich primers and paints for steel protection.

Alkyd resins: there are two categories used in film-forming polymeric coatings: drying and non-drying; both are made by condensation polymerisation of polyols such as glycerol (propane-1,2,3-triol) or pentaerythritol (2,2-bis(hydroxymethyl)1,3-propanediol) with di-carboxylic acids or anhydrides to form polyesters, either in solution with volatile organic solvents or as latex emulsions in water; inclusion of naturally derived carboxylic acid functional unsaturated oils (tall oils) creates branched polyesters with side chains that crosslink or dry on exposure to air (Figure A1.1). Alternatively, modification with acrylic and vinyl monomers, phenolic resins or polyurethanes also creates hybrid polyester alkyd coatings, which dry more quickly on solvent/ water evaporation; characterised by their good outdoor durability.

Amino resins: thermoset prepolymers formed by copolymerisation of amines such as melamine and urea with aldehydes such as formaldehyde (methanal, HCHO)—see MF resins and UF resins below—can be moulded or extruded into components, or used as crosslinkers for alkyd resins, epoxy ester resins or unsaturated acrylic or polyester resins for making coatings and adhesives.

Benzoxazine resins: formed by the reaction of phenols, formalde-hyde (methanal, HCHO) and primary amines; on heating to $\sim$200 °C/ $\sim$400 °F undergo ring-opening polymerisation to produce poly-benzoxazine thermoset networks (Figure A1.2) which, when hybrid-ised with epoxy and phenolic resins, form ternary systems with glass transition temperatures in excess of $\sim$250 °C/$\sim$490 °F; cure is char-acterised by expansion rather than shrinkage and the formation of high-molecular-weight polymers; uses include structural prepregs, liquid moulding and film adhesives for composite construction, bonding and repair where enhanced mechanical and flammability performance compared to epoxy and phenolic resins is required.

Bismaleimide (BMI) resins: formed by the reaction of maleic anhydride (furan-2,5-dione) and primary aromatic amines; chain extended polyaminobismaleimides are made by nucleophilic reaction with amine functional aromatics (Figure A1.3); BMIs contain highly reactive double bonds which on the application of heat and in the presence of catalysts undergo homopolymerisation or copolymer-isation with compounded unsaturated diluent resins, or cycloaddi-tion reactions with compounded dienes to produce thermoset crosslinked polymers with no condensation by-products; uses include the production of large-scale composite structures with dimensional stability with operating temperatures greater than 500 °F/260 °C.

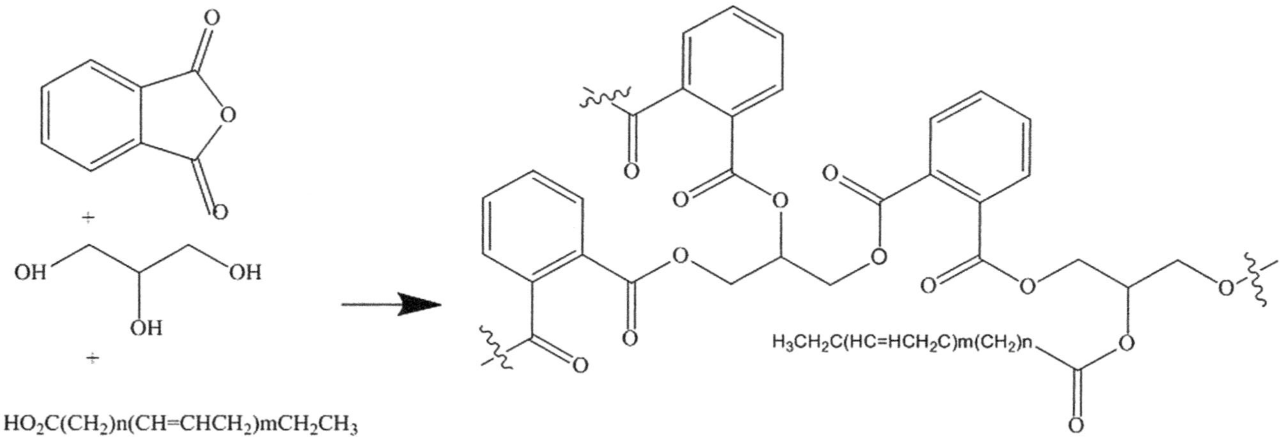

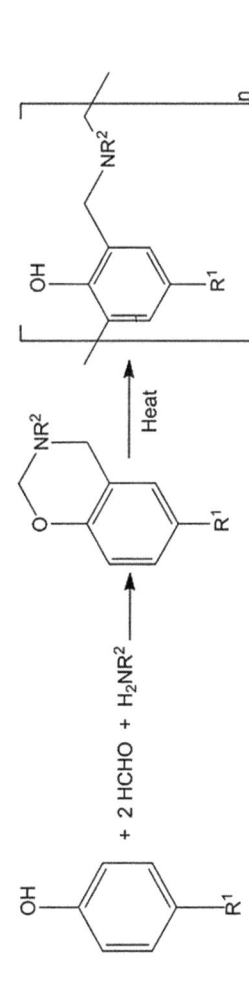

Figure A1.1 Drying alkyd resin synthetic pathway and structure.

Figure A1.2 Benzoxazine resin synthetic pathway, structure and cure mechanism.

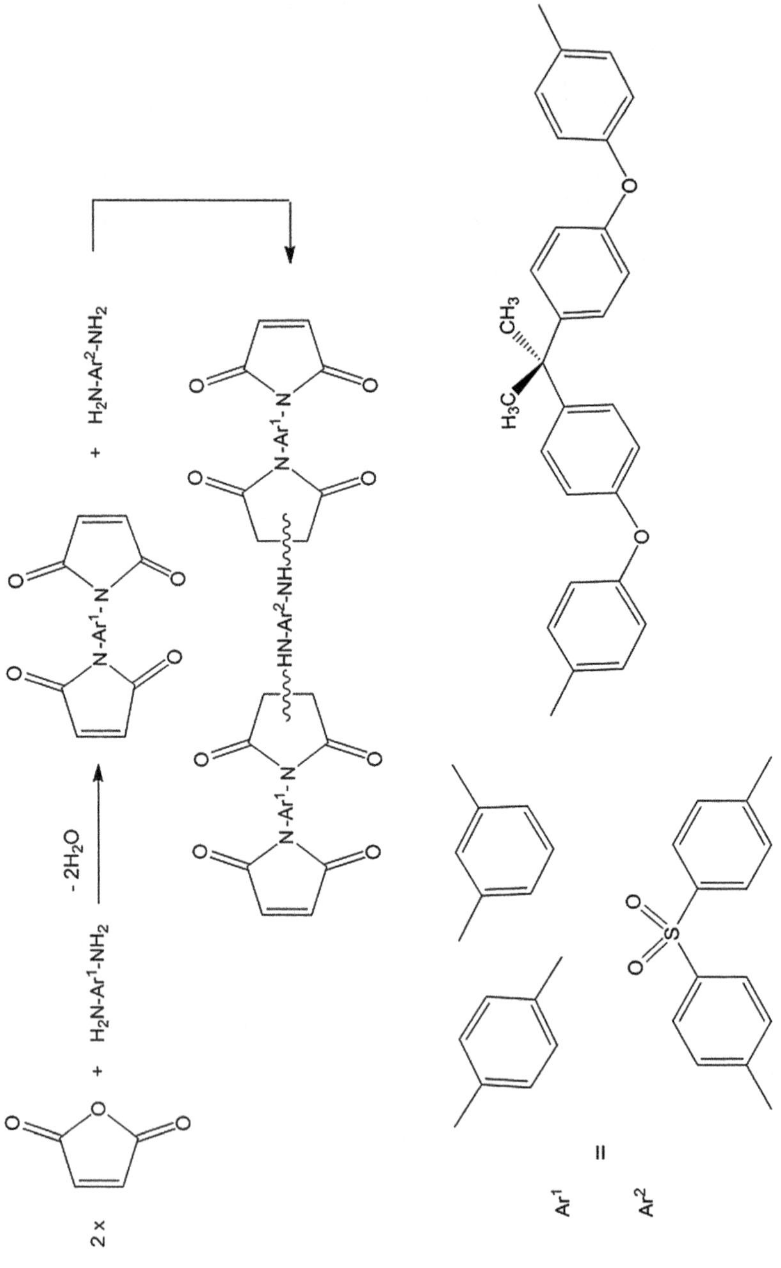

Figure A1.3 Bismaleimide resin synthetic pathway and structure.

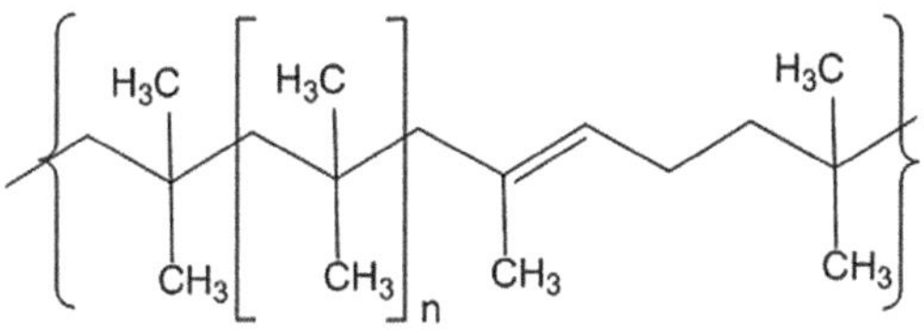

Figure A1.4 Thermoplastic butyl rubber polymer structure.

Butyl rubbers: thermoplastic copolymers made from ~98% iso-butylene ($H_2C=C(CH_3)_2$) and ~2% isoprene ($H_2C=C(CH_3)-CH=CH_2$) with long polyisobutylene segments (Figure A1.4); characterised by their tough, extensible, rubber-like properties over a wide temperature range; used to improve the durability and impermeability of adhesives, caulks, sealants and also for rubber and polymer modification.

Cyanate ester resins: –OCN (cyanate) functional resins made by the reaction of bisphenols or multifunctional phenolic novolac resins with cyanogen bromide or chloride (Br–CN or Cl–CN) to form intermediate monomers which are converted in a controlled manner into cyanate ester functional prepolymer resins by chain extension or copolymer-isation to form –OCN functional cyanurate prepolymers (Figure A1.5); during ultimate heat postcuring, all residual cyanate ester functionality polymerises by cyclotrimerisation to form tightly crosslinked poly-cyanurate networks which are characterised by their high thermal stability with glass transition temperatures up to 400 °C/752 °F and wet heat stability up to around ~200 °C/~400 °F; uses include the production of fire-, smoke- and toxicity-retardant structural composites.

Cyanoacrylate resins: formed by condensing formaldehyde (methanal) with alkyl cyanoacetates in a Knovenagel reaction; characterised by their instant moisture-initiated cure (Figure A1.6), thermoplastic nature when cured and limited temperature and chemical resistance capabilities; used for bonding all types of glass, most plastics and metal, but generally restricted to small components.

ECTFE (Ethylene-chlorotrifluoroethylene) plastics: copolymer thermoplastics made from ethylene ($H_2C=CH_2$) and chlorofluoroethylene ($F_2C=CClF$); typically semi-crystalline and applied by melt processing and powder coating to protect metals from acids even at high concentrations and temperatures, and other corrosive media including caustic, oxidizing agents and solvents.

EP and EPDM rubbers: thermoplastic copolymers of ethylene ($H_2C=CH_2$) and propylene ($H_2C=CHCH_3$) or terpolymers with diene monomers such as dicyclopentadiene (tricyclo[$5.2.1.0^{2,6}$]deca-3,8-diene) and norbornene (bicyclo[2.2.1]hept-2-ene) derivatives; used as

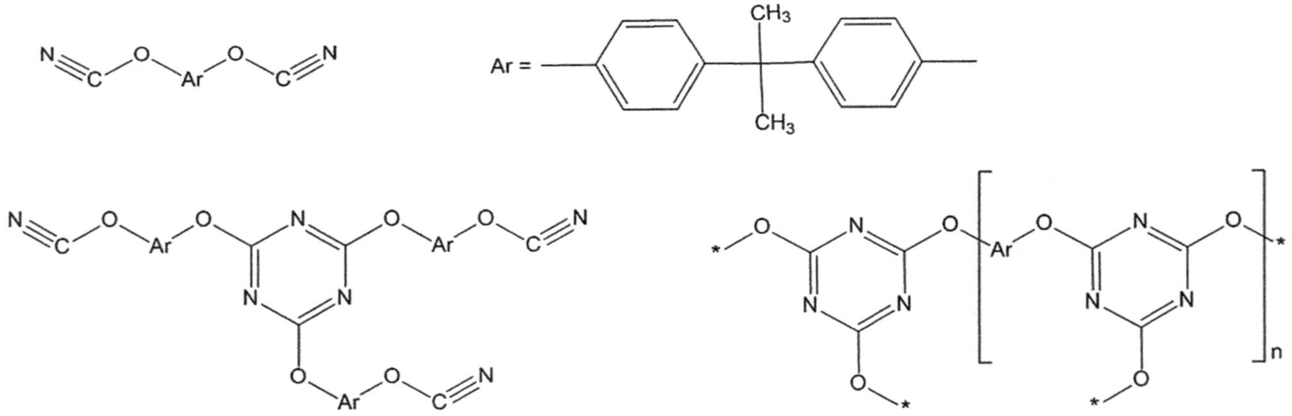

Figure A1.5 Cyanate ester monomer, prepolymer and polycyanurate structures.

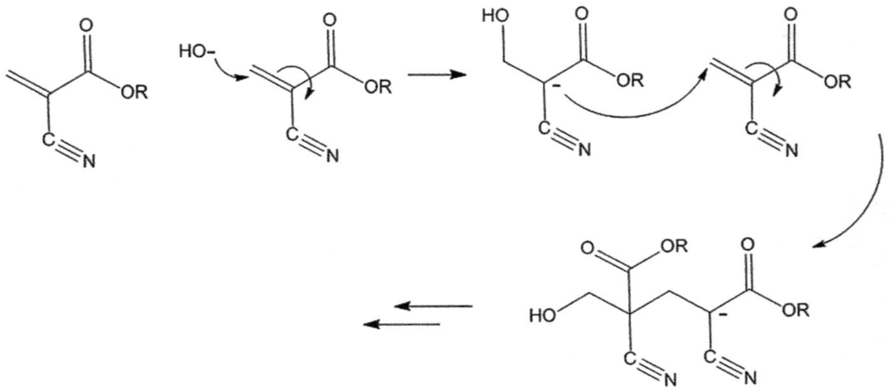

Figure A1.6 Cyanoacrylate structure and cure mechanism.

waterproofing protective sheet membranes as they retain a high level of flexibility and elasticity over a broad temperature range and are resistant to ageing; they are also used where resistance to acids, caustic solutions, ozone, sunlight and the weather is required.

Epoxy resins: thermosetting prepolymers made either by the oxidation of unsaturated cycloaliphatics or by the reaction of epichlorohydrin (chloromethyloxirane) with hydroxyl functional aromatics, cycloaliphatics and aliphatics, or amine functional aromatics—they range widely in molecular weight and physical form; the diglycidyl ethers of bisphenol-A (DGEBA) and bisphenol-F (DGEBF) are the most important class, characterised by their high adhesion, mechanical strength, heat and corrosion resistance, details of which can be found in H. Lee and K. Neville, *Handbook of Epoxy Resin,* McGraw-Hill, New York, 1967.

In the structure shown for DGEBA (Figure A1.7), "*n*" denotes repetitive subunits which range from 0 to 25; low-molecular-weight variants ($n = 0$) are liquid at ambient temperatures; higher-molecular-weight variants are more viscous liquids ($n = 1$) or solids ($n = 2+$), all types of which may be mixed with diluents, solvents and/or modifiers to meet a variety of application and performance needs; low-molecular-weight liquid epoxies through to very high-molecular-weight solid epoxies are used for a wide diversity of industrial protection, repair, restoration and bonding applications. For information on the various crosslinkers, hardeners, curing agents and catalysts used to initiate polyaddition/copolymerisation or homopolymerisation see W. R. Ashcroft, *Chemistry and Technology of Epoxy Resins,* ed. B. Ellis, Springer Netherlands, 1993, pp. 37–71, Print ISBN: 978-94-010-5302-0, Online ISBN: 978-94-011-2932-9.

Epoxy phenol novolac (EPN) resins: prepolymers made by reacting epichlorohydrin (chloromethyloxirane) with multifunctional phenolic novolac resins (Figure A1.8); there are more reactive sites compared to DGEBF resins that results in higher crosslink density thermosets; used where coatings for metal need to provide protection from corrosion, erosion or chemical attack at high continuous operating temperatures.

Epoxy ester resins: prepolymers made by reacting epoxy resins with naturally derived short, medium or long tail carboxylic acid functional fatty acid oils (Figure A1.9); used by themselves or in combination with alkyd resins as binders for ambient temperature air drying and baked-on coatings; characterised by durable and enamel-like finishes with improved chemical resistance, but less weathering resistance than straight alkyds. Epoxy resins and acrylic acid derivatives are described separately below under vinyl esters.

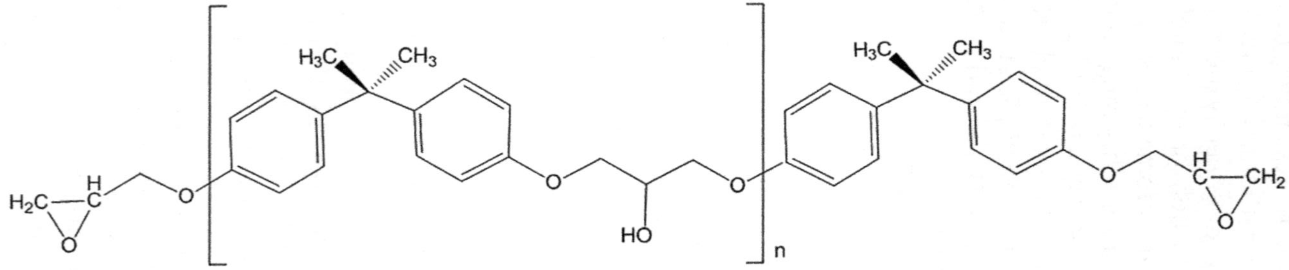

Figure A1.7 DGEBA thermoset epoxy resin structure.

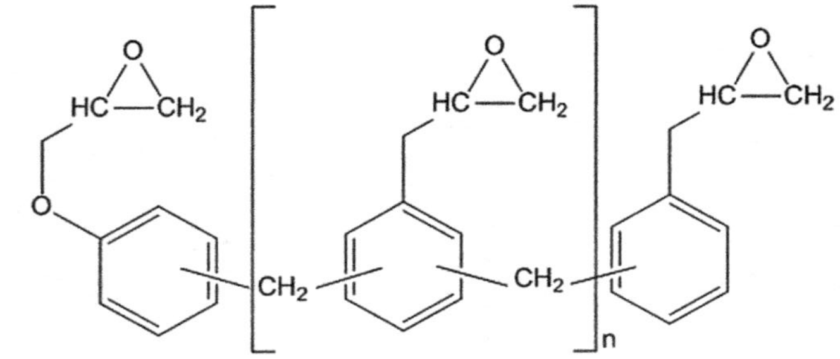

Figure A1.8 Epoxy phenol novolac thermoset resin structure.

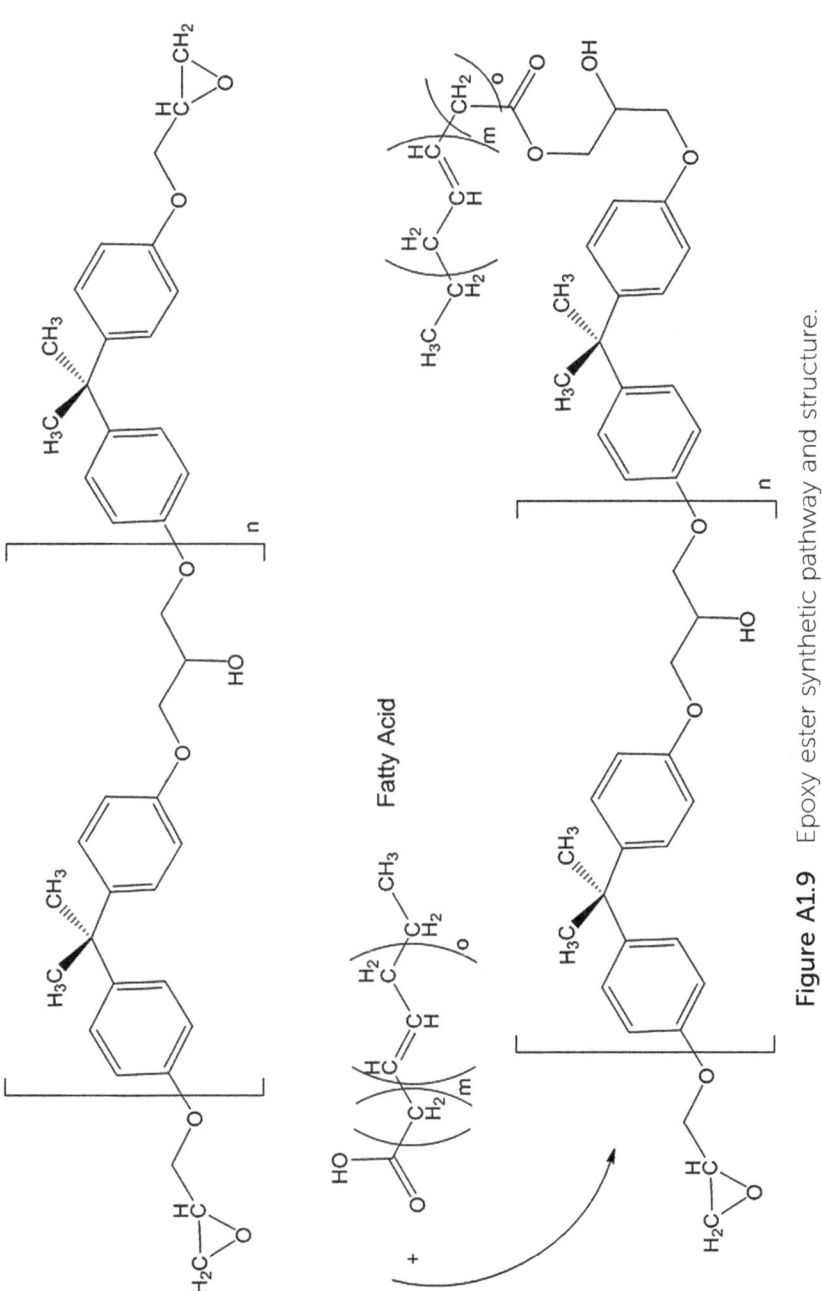

Figure A1.9 Epoxy ester synthetic pathway and structure.

$2Si(OC_2H_5)_4 + H_2O \rightarrow 2C_2H_5OH + (C_2H_5O)_3Si\text{-}O\text{-}Si(OC_2H_5)_3$, then

$(C_2H_5O)_3Si\text{-}O\text{-}Si(OC_2H_5)_3 + Si(OC_2H_5)_4 + H_2O \rightarrow 2C_2H_5OH + C_2H_5\text{-}[OSi(OC_2H_5)_2O]_3\text{-}C_2H_5$ etc.

Overall reaction: $nSi(OC_2H_5)_4 + 4nH_2O \rightarrow 4nC_2H_5OH + nSi(OH)_4 \rightarrow (SiO_2)n + 2nH_2O$

Figure A1.10 Ethyl silicate resin hydrolysis reactions.

ETFE (Ethylene-tetrafluoroethylene) plastics: copolymer thermoplastics made from ethylene ($H_2C=CH_2$) and tetrafluorethylene ($F_2C=CF_2$); characterised by high corrosion resistance and strength over a wide temperature range; preformed films typically resin bonded as liners for pipes, tanks, and vessels, where low coefficients of friction, abrasion resistance and corrosion protection are needed.

Ethyl silicate resin: also known as tetraethyl orthosilicate or tetraethoxysilane with chemical formula $Si(OC_2H_5)_4$; characterised by low viscosity and formation on hydrolysis (Figure A1.10) of precipitated silicon dioxide (silica), which firmly binds inorganic fillers and pigments to metal and other inorganic materials; used for the formulation of coating binders for zinc-rich primers and paints for steel protection.

EVA (Ethylene-vinyl acetate) plastics: thermoplastic copolymers made primarily (60–90%) from ethylene ($H_2C=CH_2$) and lesser amounts (10–40%) of vinyl acetate (ethenyl acetate, $H_2C=CHOCOCH_3$); high levels of ethylene are incorporated into the polymer backbone for softness and hydrophobicity, low levels of vinyl acetate are used to provide hardness and hydrophilicity; characterised by their overall flexibility, impact and puncture resistance; used on concrete as high wear-resistant waterproofing membranes.

Furan resins: thermosetting prepolymers formed from furfuryl alcohol (2-furanmethanol) (Figure A1.11), or by modification of furfural (furan-2-carbaldehyde) with phenol, formaldehyde, urea or other extenders; characterised by their high resistance to heat, acids and alkalies; formulated as single-component elevated temperature (baking) systems with or without latent acid catalysts, or as dual-component no-bake acid-hardened systems for use as cements, adhesives, casting resins, coatings and composites.

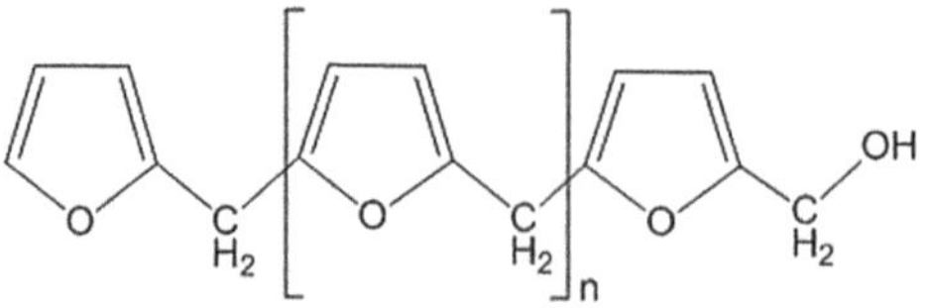

Figure A1.11 Polyfurfuryl alcohol furan resin

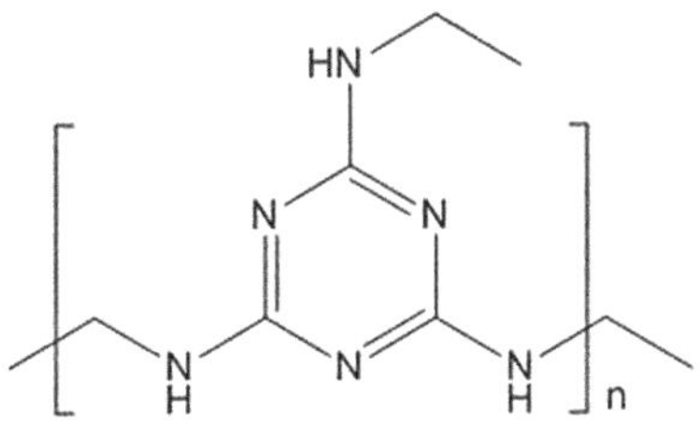

Figure A1.12   Cured melamine resin.

FEP (Fluorinated ethylene-propylene) plastics: copolymer thermoplastics made from hexafluoropropylene ($F_2C=CF(CF_3)$) and tetrafluoroethylene ($F_2C=CF_2$); characterised by chemical inertness and low friction properties; used as liners for pipes, tanks and vessels.

MF (Melamine-formaldehyde) resins: thermoset prepolymers formed by polymerisation of melamine (1,3,5-Triazine-2,4,6-triamine) with formaldehyde (methanal, HCHO); characterised by rapid cure under the influence of heat, catalysts and pressure, resulting in tightly crosslinked three-dimensional crystalline-like rigid polymers (Figure A1.12); used for plastic laminating, and in butylated and solvented form as crosslinkers with alkyd resins, epoxy ester resins and unsaturated acrylic or polyester resins for surface coating applications.

MMA resins: linear polymethylmethacrylate ($-CH_2-C(CH_3)(CO_2CH_3)-)_n$ blends dissolved in monomeric methyl methacrylate ($H_2C=C(CH_3)COOCH_3$)—also known as acrylic resins (see above); characterised by rapid thermoset cure rates, toughness, strength and hydrophobicity; used in the formulation of fast thermosetting adhesives, flooring and waterproofing and wearing layers.

NBR (Nitrile-butadiene) or NR rubbers: thermoplastic copolymers of acrylonitrile (2-propenenitrile, $H_2C=CHCN$) and buta-1,3-diene ($H_2C=CH-CH=CH_2$); characterised by flexibility and resistance to oil, fuel and other chemicals—the lower the nitrile content, the higher the flexibility, the higher the nitrile content, the higher the resistance to oils; used for the manufacture of oil-resistant rubber components and also as modifiers for epoxy and phenolic resins.

PA (Polyamide) plastics: thermoplastic semi-crystalline polymers made by poly-condensation of diamines with dicarboxylic acids, for example, PA66 (Nylon 66) from hexane-1,6-diamine ($H_2N(CH_2)_6NH_2$) with adipic acid (hexanedioic acid, $HOCO(CH_2)_4COOH$), or by self-polymerisation of lactams or amino acids such as PA6 (Nylon 6) from ε-caprolactam (azepan-2-one) or PA12 (Nylon 12) made from ω-aminolauric acid (12-aminododecanoic acid) or laurolactam (azacyclotridecan-2-one); anionic PA-6 (APA-6) is used like a thermoset

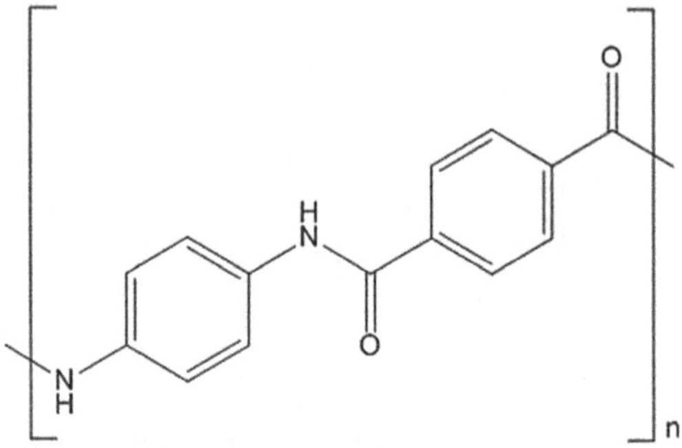

Figure A1.13 Aramid polyamide thermoplastic polymer structure.

in the manufacture of fibre reinforced polymer matrix composites as it converts rapidly into highly crystalline, tough, durable and inert PA-6; aromatic polyamides (Aramids) such as Kevlar® and Twaron®, made from 1,4-diaminobenzene and terephthaloyl chloride, are highly oriented thermoplastic polymers (Figure A1.13) with high heat resistance and strength and can be spun into fibres for use as reinforcement in polymer matrix composites.

Polyaspartic coating resins: polyaspartic aliphatic polyureas, known simply as polyaspartics, are produced in thermoset reactions between aspartic acid esters (diesters of 2-aminobutanedioic acid) and aliphatic polyisocyanates; characterised by UV and light stability, abrasion resistance and chemical resistance; used to formulate high solids (based on solid isocyanates) and 100% solids (based on fluid polyisocyanates) non-yellowing decorative finish concrete coatings and sealers.

PC (Polycarbonate) plastics: amorphous thermoplastic polymers most frequently formed by polycondensation of bisphenols such as bisphenol-A with carbonyl dichloride ($COCl_2$) (Figure A1.14); characterised by optical transparency, stiffness and toughness which are retained over a wide temperature range (-20 °C/ -4 °F up to $\sim$140 °C/ $\sim$280 °F); also distinguished for their ease of working through moulding and thermoforming; one of the most widely used engineering thermoplastics.

PE (Polyester) plastics: thermoplastic linear polymers formed by polycondensation of aromatic diacids such as terephthalic acid ($C_6H_4(COOH)_2$, benzene-1,4-dicarboxylic acid) or isophthalic acid (benzene-1,3-dicarboxylic acid) with aliphatic diols such as ethane-1,2-diol, propane-1,3-diol, butane-1,4-diol, or cycloaliphatic diols such

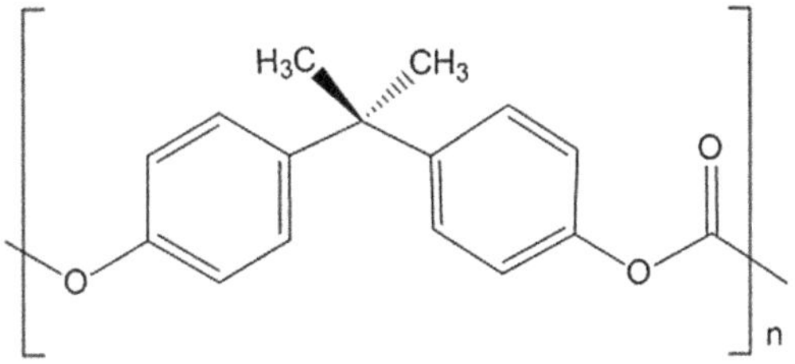

Figure A1.14 Polycarbonate thermoplastic polymer structure.

as cyclohexane-1,4-dimethanol (Figure A1.15); characterised by high degrees of crystallinity, hardness, dimensional stability and chemical resistance; used for blow moulding and extrusion of engineering components, film and fibres.

PE (Polyethylene) plastics: of the various grades of thermoplastic polymers made from ethylene ($H_2C=CH_2$), only the ultra-high-molecular-weight polyethylenes (UHMWPEs) with molecular weights between 3.1 and 5.67 million Daltons are used in engineering component manufacture or as a fibre reinforcement in composites; characterised by toughness and resistance to chemicals including strong acids and bases, gentle oxidants and reducing agents.

PEEK (Polyether ether ketone): the best known of the thermoplastic semi-crystalline polyaryletherketones (Figure A1.16); made by step-growth polymerisation of activated aryl dihalides with aromatic diphenolates or by Friedel-Crafts acylation; characterised by their dimensional stability, chemical and solvent resistance and thermal stability—composites with a PEEK matrix can have a continuous use temperature up to 250 °C/480 °F.

PEI (Polyether imide): amorphous thermoplastic polymers formed by polycondensation of aromatic diamines with aromatic ether backboned dianhydrides (Figure A1.17); characterised by stiffness and strength, broad-range chemical resistance and good thermal properties with a T_g of 215 °C/419 °F and a continuous use temperature of $\sim$170 °C/$\sim$340 °F; used to make structural composites by a wide variety of processing methods.

PF (Phenol-formaldehyde) resins: thermosetting resins, which are either novolacs or resoles; **Novolacs** are made with acid catalysts and a molar ratio of formaldehyde to phenol of less than one to give methylene linked phenolic oligomers (Figure A1.18); **Resoles** are made with alkali catalysts and a molar ratio of formaldehyde to phenol of greater than one to give hydroxymethyl terminated phenolic

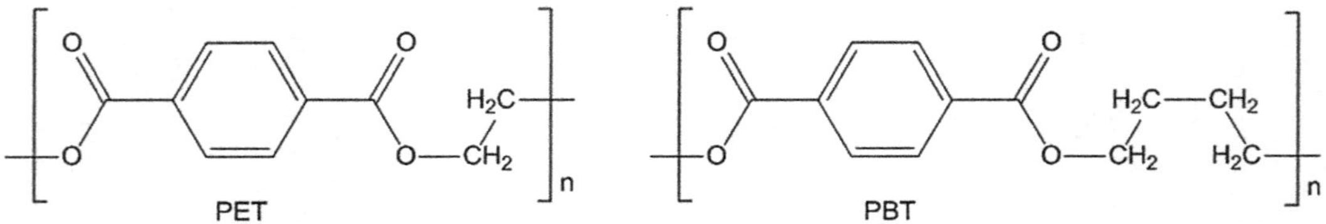

Figure A1.15 PET and PBT polyester thermoplastic polymer structure.

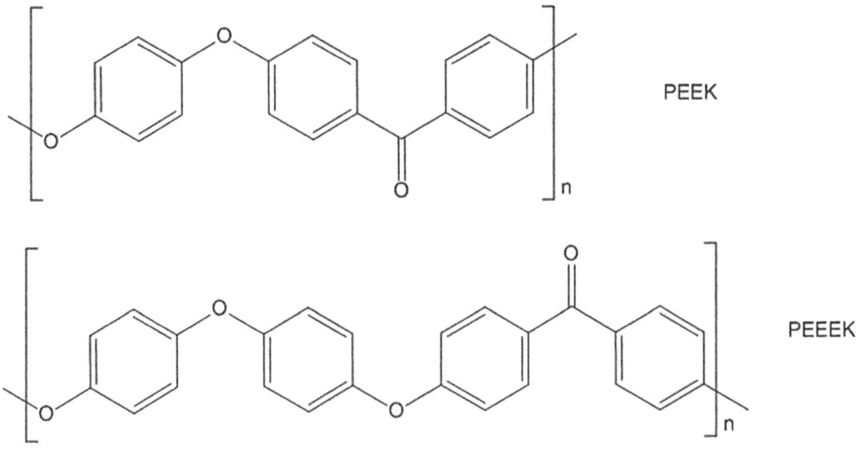

Figure A1.16   Polyarylether ketone thermoplastic polymer structures.

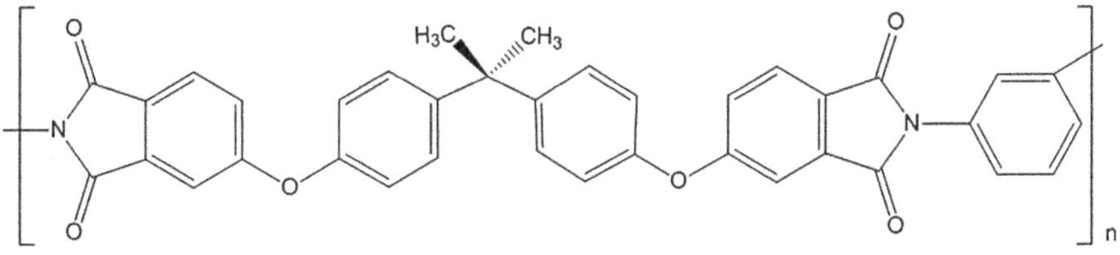

Figure A1.17   Polyether imide thermoplastic polymer structure.

oligomers with methylene and benzylic ether-linked phenol units (Figure A1.19); both types are characterised by high bonding strength and crosslink density capabilities; used in high-temperature adhesive, coating and moulding applications with various fillers, reinforcements, catalysts or co-curing resins such as epoxies.

PFA (Perfluoroalkoxy alkane) plastics: copolymer thermoplastics made from tetrafluoroethylene ($F_2C{=}CF_2$) and perfluorovinyl ethers ($F_2C{=}CFOCF_3$); characterised by smaller chain length and higher chain entanglement than ECTFE, ETFE, FEP, PTFE and PVDF; used where resistance to flexural stress is required at high temperatures.

PI (Polyimide): can be thermoplastic or thermoset; thermoplastic types are linear and made by reacting aromatic dianhydrides with an aromatic diamine or diisocyanate—Kapton® (Figure A1.20), made from 4,4′-diaminodiphenyl ether and pyromellitic dianhydride, is flexible and stable across a wide temperature range from −269 to +400 °C (−452 to 752 °F); thermoset polyimides, typically used for prepregs, are amide-acid prepolymers with varying thermoset endcap

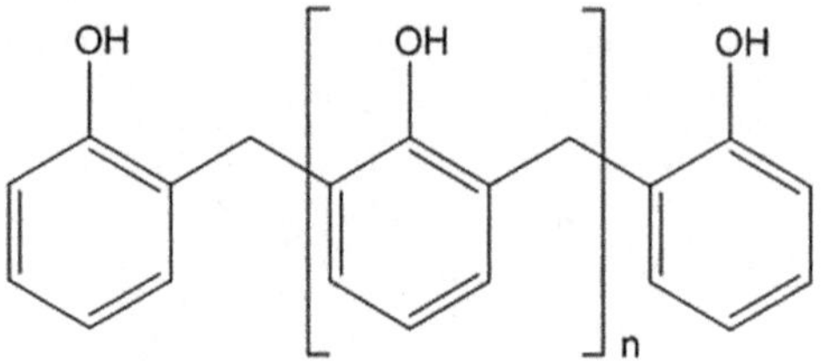

Figure A1.18 Novolac phenolic resin structure.

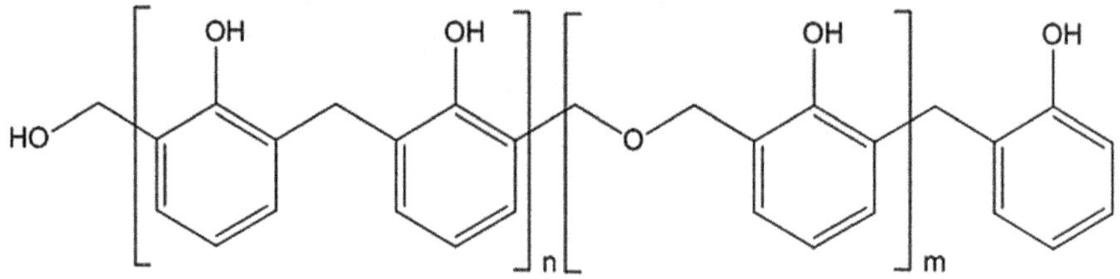

Figure A1.19 Resole phenolic resin structure.

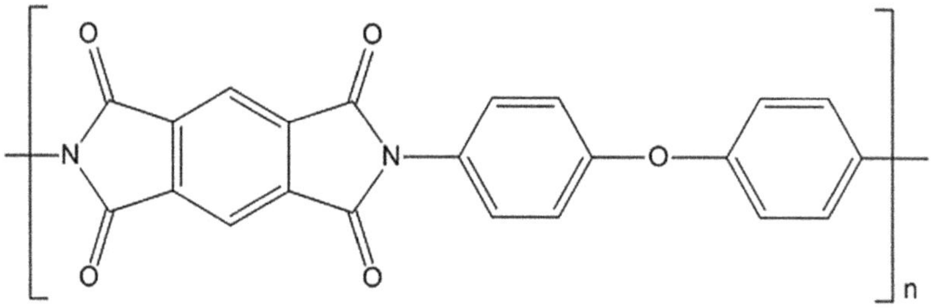

Figure A1.20 Polyimide thermoplastic polymer structure.

functionality types characterised by higher thermo-oxidative stability with continuous use at temperatures up to ~450 °C/~840 °F. For more details, see D. A. Scola and J. H. Vontel, *Polym. Compos.*, 1988, **9**(6), 443–452.

PMMA resins: see MMA resins above.

PP (Polypropylene) plastics: thermoplastic semi-crystalline isotactic polymers with all methyl groups on the same side of the polymer chain (Figure A1.21) made by homopolymerisation of propylene ($H_2C=CHCH_3$); characterised by light weight and toughness; used as the polymer matrix for glass fibre reinforced thermoplastic structural composite component manufacture.

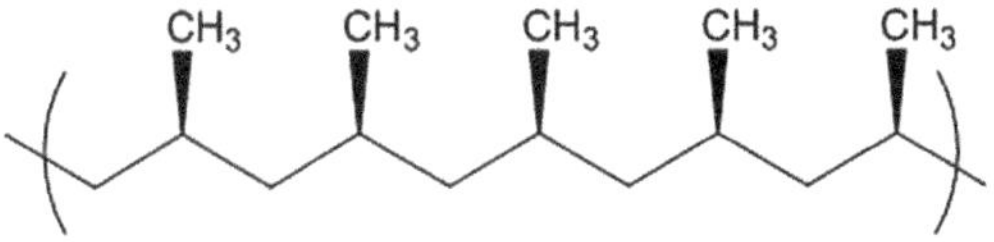

Figure A1.21 Polypropylene thermoplastic polymer structure.

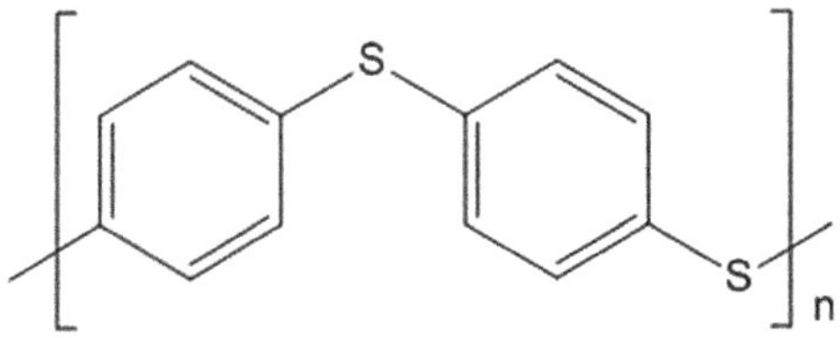

Figure A1.22 Polyphenylene sulfide thermoplastic polymer structure.

PPS (Polyphenylene sulfide) plastic: thermoplastic semi-crystalline polymer with a melting point of ~285 °C/~545 °F (Figure A1.22); characterised by dimensional stability, heat resistance, chemical and biological attack resistance, flame retardency; used in combination with fibres and fillers to overcome brittleness for moulded and extruded advanced engineering component fabrication.

Polysiloxane polymers: inorganic polymers with multiple siloxane (Si–O–Si) linkages which are resistant to oxidation, heat and ultraviolet degradation; formed from silicone resins (see below) or functionalised polysiloxanes—amine functional polysiloxanes are used as cross-linkers for hybrid urethane, epoxy, acrylic, polyester and alkyd, acrylic coatings—acrylate, vinyl ether and epoxy functional polysiloxanes are used to create thermoset, UV or electron beam cured composites, coatings and adhesives with enhanced temperature resistance.

Polysulfide resins: liquid polysulfide backboned hydrocarbon polymers with terminal thiol functionality such as $HS(CH_2CH_2O$ $CH_2OCH_2CH_2-S-S)_n-CH_2CH_2OCH_2OCH_2CH_2-SH$; characterised by flexibility, impact resistance and chemical/oil resistance; used as base polymer with inorganic and peroxide catalysts to make high bond strength, flexible and durable sealants, or as modifiers for epoxy and polyurethane resins for the formulation of sealants, adhesives and protective coatings; epoxy-terminated polysulfides also available for use alone or in combination with other epoxy resins.

PTFE (Polytetrafluoroethylene) plastic: thermoplastic homo-polymers made from tetrafluorethylene ($F_2C=CF_2$); characterised by high strength, toughness, flexibility and self-lubrication at temperatures between minus ~200 °C/minus ~328 °F and +260 °C/+500 °F;

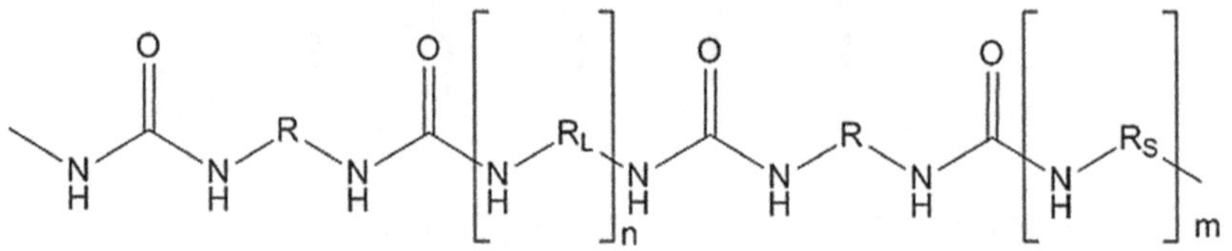

Figure A1.23   Polyurea thermoset polymer structure.

used to line pipes, tanks and vessels where low coefficients of friction, abrasion resistance and corrosion protection are needed.

Polyureas: thermoset elastomeric polymers with urea (–NH–CO–NH–) linkages made by combining diisocyanates (OCN–R–NCO), which can be aliphatic or aromatic in nature and either monomeric or isocyanate-terminated prepolymers, with blends of long-chain amine-terminated polyether or polyester resins ($H_2N-R^L-NH_2$) and short-chain diamine extenders ($H_2N-R^S-NH_2$) (Figure A1.23); characterised by near instantaneous cure, good adhesion to metal, concrete or plastics, and also high tensile strength and resistance to corrosion and wear; typically formulated with $1:1$ volume mix ratios to facilitate spray application as rapid curing protective and abrasion resistant coatings.

PU (Polyurethane) resins: thermoset prepolymers with carbamate (–NH–CO–O–) links which are linear and elastomeric if formed by combining diisocyanates ($OCN-R^1-NCO$) with long chain diols ($HO-R^2-OH$) (Figure A1.24); or crosslinked and rigid if formed from polyisocyanates and hydroxyl-terminated polyether or other resin types with an average of at least two hydroxyl groups per molecule; characterised by high adhesion, flexibility and resistance to fatigue; can be solid or have an open cellular structure if foamed; solid linear and crosslinked PUs are used for elastomers, sealants, adhesives and protective coatings, where resistance to abrasion and corrosion are required. For more information on the wide variety of building blocks available and polymer structures achievable with solid and foamed thermoset polyurethane resins, see *Polyurethane Handbook*, ed. G. Oertel, Hanser, Munich, Germany, 2nd edn, 1994, ISBN-10: 1569901570; ISBN-13: 978-1569901571.

PVA (Polyvinyl acetate) polymers: thermoplastic emulsion homo-polymers made from vinyl acetate (ethenyl acetate, $CH_3CO_2CH=CH_2$); used as the film-forming ingredient in water-based paints, polymer modified cements and adhesives as they partially hydrolyse in service to water-soluble polyvinyl alcohol.

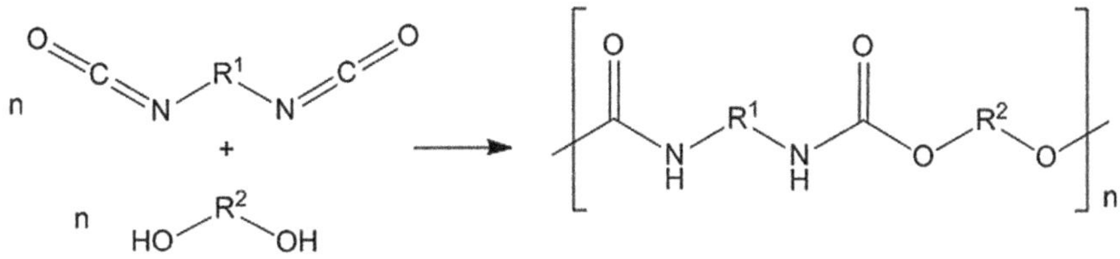

Figure A1.24 Linear polyurethane thermoset polymer structure.

PVAC (Polyvinyl acrylic) polymers: emulsion copolymers made from vinyl acetate with acrylic and methacrylic acids and their esters; distinguished by wet tack strength, adhesion, wet and dry tensile strength and ability to form tough and flexible films with or without the use of coalescing solvents or glycols; used in adhesives, architectural/masonry coatings and liquid-applied roof waterproofing.

PVC (Polyvinyl chloride) membranes: thermoplastic homopolymers made from vinyl chloride (chloroethene, $H_2C=CHCl$) and compounded with plasticisers to improve flexibility, extensibility and resilience for use as corrosion and weathering-resistant waterproofing sheet membranes.

PVDF (Polyvinylidene fluoride) plastics: thermoplastic semi-crystalline homopolymers made from vinylidene fluoride (1,1-difluoroethene, $F_2C=CH_2)C=CFOCF_3$); characterised by thermal stability and resistance to chemicals, they are used for industrial linings/barrier coatings.

Silicone resins: liquid silicones (polyorganosiloxanes, $(R_2SiO)_n$) in which silicon is directly bonded to carbon and there is at least one oxygen atom bonded to the silicon; physical form (oil, grease, rubber, plastic) and use varies with molecular weight, structure (linear, branched) and nature of substituent groups (R = alkyl, aryl, H, OH, alkoxy).

- High-molecular-weight linear polydimethylsiloxanes $(R = CH_3)$ are classed as silicone oils, not resins, and are used primarily as lubricants, hydraulic fluids and heat resistant oils;
- Silicone resins used to form hydrophobic three-dimensional silicate resin networks within the pores of concrete and masonry, or as binders for silicone resin emulsion paints and plasters, are silane (R = H) and siloxane (R = alkyl) functional.
- Branched and cage-like organosiloxane resins with functional substituents (R = H, OH, and/or alkoxy) self-condense to form highly crosslinked, insoluble polysiloxane polymer networks that

can be used by themselves as binders for highly heat resistant coatings, or to enhance the thermal stability of alkyd, acrylic, UPR, VE or epoxy coatings;

– Phenyl substituted (R = Ph) silicone resins have greater thermal stability than methyl substituted (R = CH$_3$) silicone resins when polymerised at temperatures between ~150 °C (~300 °F) and ~200 °C (~400 °F)—at temperatures above ~300 °C (~600 °F) all organic substituents pyrolytically decompose leaving crystalline silicate polymers of stoichiometric composition (–SiO$_2$–)$_n$.

SA (Styrene-acrylic) resins: copolymer emulsions made from styrene (phenylethene, C$_6$H$_5$CH=CH$_2$) and one or more substituted (meth)acrylate ester monomers (H$_2$C=C(R^1)COOR^2), which introduce side-chain crosslinkable functionality through the R^2 ester groups; characterised by adhesion, water resistance and flexibility, they are used with or without external crosslinking agents as film-forming binders for masonry coatings and polymer modified cements.

SBC (Styrene block copolymer) resins: linear thermoplastic copolymers such as styrene-butadiene-styrene (SBS), styrene-ethylene-butadiene-styrene (SEBS), styrene-ethylene-propylene-styrene (SEPS) or styrene-isoprene-styrene (SIS); properties depend on the molecular weight of the polystyrene end-blocks that contribute hardness and strength, the chain lengths of the rubbery mid-block polymers which confer tack and adhesion, and also on the diblock/triblock ratio; used with tackifiers and plasticisers in diverse adhesive applications, and also for liquid applied roof waterproofing membranes.

SBR (Styrene-butadiene rubber) resins: emulsion polymerised random copolymers of styrene (phenylethene, C$_6$H$_5$CH=CH$_2$) and buta-1,3-diene (H$_2$C=CH—CH=CH$_2$); characterised by their residual backbone unsaturation (Figure A1.25), broad molecular weight distributions, high degree of branching and nanoparticle size; used to increase adhesion, flexibility, crack resistance and improve water impermeability of masonry coatings, polymer concretes and liquid asphalts.

TPU (Thermoplastic polyurethane) plastics: linear segmented block copolymers with hard and soft segments which impart thermoplastic and elastomeric behaviour; produced by the reaction of diisocyanates with diols—the hard block rigid segments being made from aromatic or aliphatic diisocyanates combined with short-chain diols—the soft block elastomeric segments created through the incorporation of

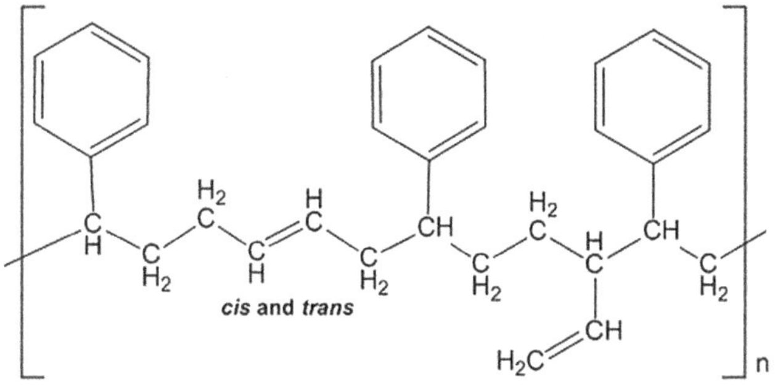

Figure A1.25 Styrene-butadiene thermoplastic polymer structure.

long-chain polyether or polyester diols; characterised by multiple intermolecular hydrogen bonds between urethane linkages in hard block segments which leads to aggregation, order and the formation of crystalline or pseudo crystalline areas within a soft and flexible soft block segment matrix; used for a variety of extruded film, sheet and profile engineering applications.

UF (Urea-formaldehyde) resins: thermosetting prepolymers made from urea (carbamide, H_2NCONH_2) and formaldehyde (methanal, HCHO) which are mostly low-molecular-weight condensates (Figure A1.26); characterised by rapid cure under the influence of heat, catalysts and pressure resulting in tightly crosslinked three-dimensional crystalline-like rigid polymers; used primarily as adhesives for the bonding of plywood, particleboard and other engineered wood components.

UPR—Unsaturated polyester resins: thermoset condensation polymers formed by the reaction of mixtures of glycols with unsaturated dibasic acids or anhydrides with or without aromatic dibasic acids or anhydrides—propylene glycol (propane-1,2-diol) is the chief glycol used in blends with ethylene glycol, diethylene glycol or neopentyl glycol—maleic anhydride (furan-2,5-dione) typically provides critical backbone unsaturation needed for crosslinking (Figure A1.27); characterised by versatility in mechanical properties and chemical resistance, depending on the choice of monomers; inclusion of orthophthalic anhydride (2-benzofuran-1,3-dione, $C_6H_4(CO)_2O$) is used to make standard orthophthalic polyester resins whereas isophthalic acid (benzene-1,3-dicarboxylic acid) or terephthalic acid (benzene-1,4-dicarboxylic acid) are used to make more specialised isophthalic/terephthalic polyester resins, where superior structural and corrosion properties are required; additive monomers such as styrene or vinyl

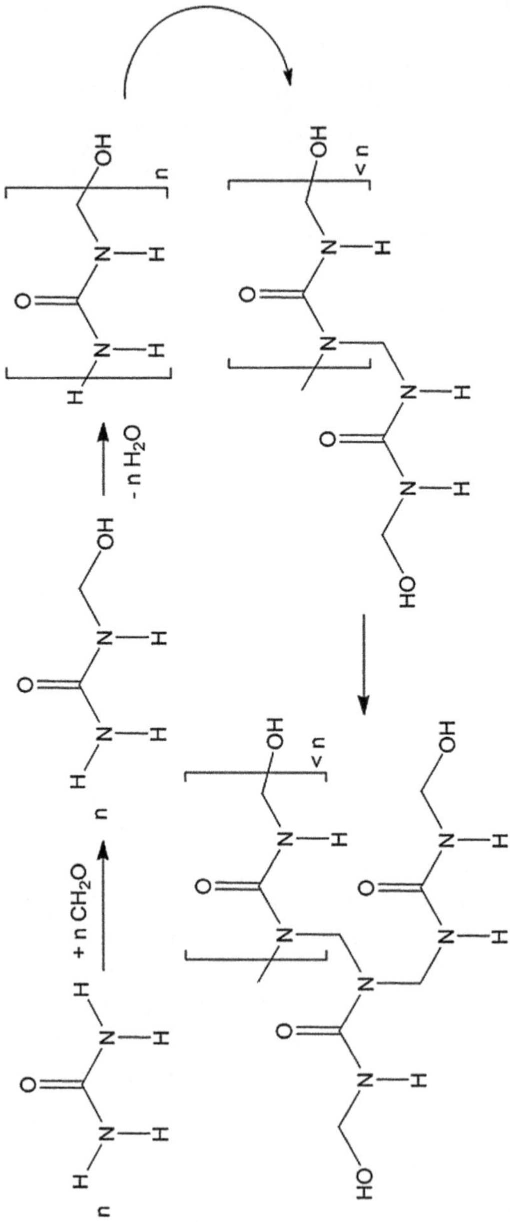

Figure A1.26 Linear and branched urea-formaldehyde resin condensate structures.

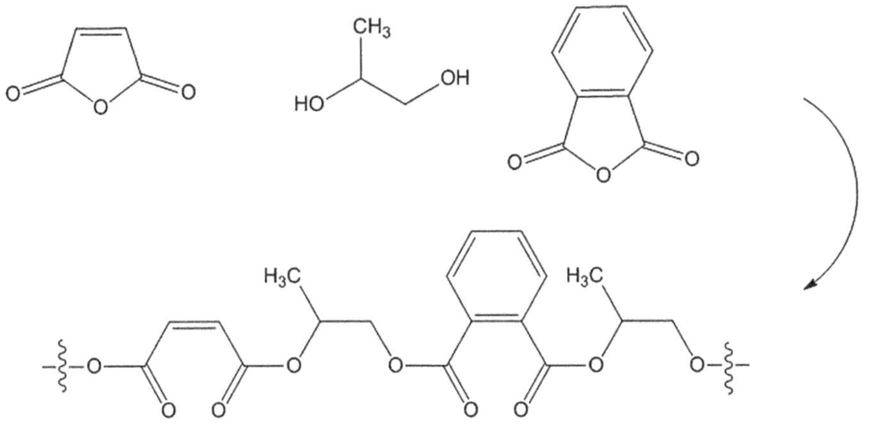

Figure A1.27 Idealised chemical structure for orthophthalic polyester made from maleic anhydride, propylene glycol and orthophthalic anhydride.

acetate are used to lower the viscosity to enhance ease of application and to provide links between polymer chains when free radical initiator catalysts are mixed in at time of application; used for laminating and in other composite manufacturing processes, as well as for formulating fast-curing repair compounds, primers and coatings.

VAC (Vinyl acrylic) resins: see PVAC polymers above.

VAE (Vinyl acetate ethylene) dispersions: thermoplastic copolymer emulsions made primarily (60–95%) from vinyl acetate (ethenyl acetate, $H_2C=CHOCOCH_3$) and lesser amounts (5–40%) of ethylene ($H_2C=CH_2$); distinguished by flexibility and adhesion in low temperature and wet conditions; used for architectural/masonry coatings, as cement modifiers, and for construction adhesives.

VE (Vinyl ester) resins: thermosetting unsaturated prepolymers made (Figure A1.28) by reacting epoxy functional resins with acrylic acid derivatives such as methacrylic acid (2-methylpropenoic acid, $H_2C=C(CH_3)CO_2H$), crotonic acid ((E)-2-butenoic acid, $CH_3CH=CHCO_2H$) or cinnamic acid ((E)-3-phenylprop-2-enoic acid, $C_6H_5CH=CHCO_2H$); characterised by adhesion, heat resistance and corrosion resistance and being typically stronger than polyesters and more resilient to impact than epoxies; supplied in reactive solvents such as styrene and used with catalysts to formulate fast setting industrial protection, repair and restoration materials.

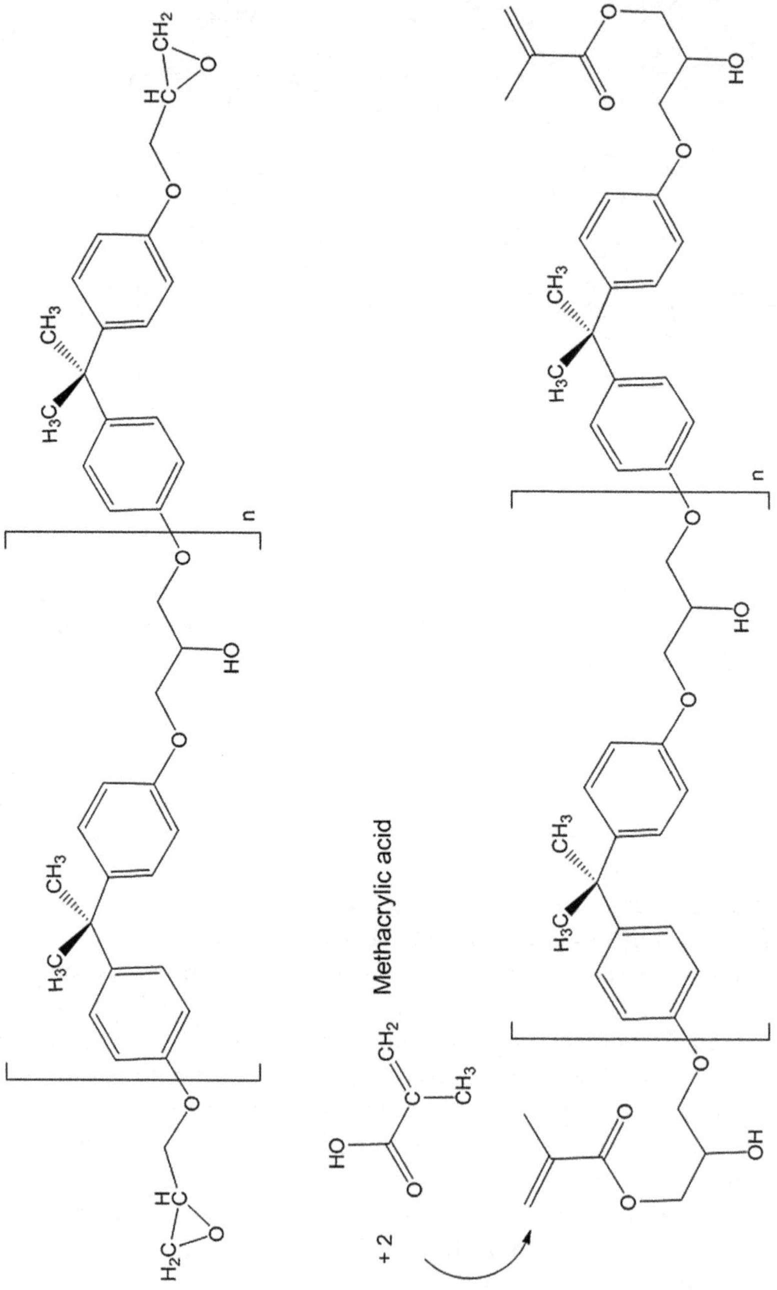

Figure A1.28 Vinyl ester synthetic pathway and structure.

Appendix 2 Thermoset Resin Stoichiometry Calculations

A2.1 Unsaturated Polyester, Vinyl ester, and Acrylic Resins

Unsaturated carbon–carbon double bonds at the ends or on the backbone of unsaturated polyesters, vinyl esters and acrylic resins will copolymerise with unsaturated co-monomer diluents such as styrene, vinyl ether or methyl methacrylate monomer to form crosslinked networks in reactions accompanied by an exothermic release of energy. Polymerisation is initiated by free radicals generated from ionizing radiation or by the photolytic or thermal decomposition of a radical initiator.

In the case of electron beam radiation, the energy is sufficient for free radicals to be formed directly from the resins and monomers, whereas visible light and UV radiation requires photo-initiators, which decompose to form free radicals and/or cations.

Conventional two-part thermoset unsaturated polyesters, vinyl esters and acrylics are formulated with initiators that are affected by heat or by auxiliary accelerators which promote decomposition to form radicals. The speed of cure of unsaturated polyesters, vinyl esters and acrylic resins is principally controlled by the amount of accelerator included or by the irradiation energy exposure, and the heat of reaction in thick sections can also hasten development of crosslinking and mechanical property development.

Industrial Polymer Applications: Essential Chemistry and Technology
By William R. Ashcroft
© William R. Ashcroft 2017
Published by the Royal Society of Chemistry, www.rsc.org

As such, and unlike epoxies and polyurethanes, there are no stoichiometric considerations for free radical initiated copolymerisation crosslinking—the type and proportions of unsaturated resin, unsaturated monomer diluent and initiators/accelerators where needed are selected empirically based on the ultimate physical, mechanical and chemical resistance properties required as well as the intended method of application/fabrication and exotherm control constraints.

A2.2   Furan, Phenolic and Amino Resins

Furan resin prepolymers made from furfuryl alcohol, or by modification of furfural with phenol, formaldehyde, urea or other extenders, are analogous to phenolic (phenol-formaldehyde) and amino (urea- and melamine-formaldehyde) thermosetting resins in that they cure by polycondensation involving release of water as well as heat. Furan resins and furan hybrid resins can be formulated as single-component elevated temperature (baking) systems with or without latent acid catalysts, or as dual component (no-bake) acid-hardened systems, whereas phenolic and amino resins cure under the influence of heat, catalysts and pressure.

In practice, there are no fundamental stoichiometric considerations for baked or non-baked acid-catalysed thermoset condensation of furan, phenolic, amino or hybrids and mixtures of the three resin classes, or when used to crosslink alkyd resins, epoxy ester resins or unsaturated acrylic or polyester resins, even though for every hydroxymethyl or alkoxymethyl group there is one crosslinking point established for each mole of water or alcohol released. The degree of pre-polymerisation and level of etherification/residual hydroxymethyl content in the resins, the catalyst selection/loading, the curing temperature and method of application all influence the ability to control polymerisation exotherm, the cure rate and the ultimate cured properties which are correspondingly balanced and determined empirically.

A2.3   Liquid Epoxy, Epoxy Phenol Novolac and Epoxy Ester Resins

Liquid epoxies can be cured by catalytic homopolymerisation in the presence of anionic or cationic catalyst types, but they are more commonly copolymerised with stoichiometric or near-stoichiometric quantities of multi-functional amine based curing agents/cross-linkers/hardeners through polyaddition reactions to achieve maximum mechanical properties and resistance to chemical corrosion. Catalysts can be used to accelerate nucleophilic addition reaction rates, but the selection of the type and functionality of the amine curative as well as the epoxy resin from the various aromatic, araliphatic, aliphatic and cycloaliphatic options available also has a strong influence on cure rates in addition to ultimate cured physical, mechanical and chemical resistance properties.

A2.3.1 Basic Stoichiometry Calculations

The Epoxide Equivalent Weight (EEW), known alternately as Weight Per Epoxy Equivalent (WPE), for an epoxy resin or diluent of known molecular weight and functionality is determined from eqn (A2.3.1).

$$\text{Epoxide Equivalent Weight (EEW)} = \frac{\text{Molecular Weight of Epoxy Resin}}{\text{Number of Epoxy Groups}}$$

$$(A2.3.1)$$

In practice, however, commercial diglycidyl ether resins consist of a distribution of molecular weights rather than a single idealised structure so their epoxide content, also known as epoxide number, is determined experimentally by ASTM D1652 titrimetric analysis. Epoxide number signifies the number of epoxide equivalents found in 1 kg of resin or, when expressed as EEW (eqn (A2.3.2)), the weight of resin containing 1 mole equivalent of epoxide (g mol^{-1}).

$$\text{Epoxide Equivalent Weight (g mol}^{-1}) = 1000/\text{Epoxide Number (eq. kg}^{-1})$$

$$(A2.3.2)$$

The stoichiometric quantity of an amine hardener required to solidify an epoxy resin is typically designated as the loading in Parts per Hundred by weight of Resin, abbreviated to PHR, and calculated from eqn (A2.3.3).

$$\text{Loading in PHR of amine hardener} = \frac{\text{AHEW} \times 100}{\text{EEW}} \qquad (A2.3.3)$$

where Amine Hydrogen Equivalent Weight (AHEW) of an amine hardener of known molecular weight and functionality is calculated in turn from eqn (A2.3.4)

$$\text{AHEW} = \frac{\text{Molecular Weight of Amine}}{\text{Number of Active Hydrogens}} \qquad (A2.3.4)$$

Where other types of curing agents/crosslinkers/hardeners are utilized, such as polymeric amines, carboxylic acids and anhydrides, phenol formaldehyde resins, amino formaldehyde resins, polysulfides and polymercaptans, loadings are calculated from the active equivalent weights where it is possible to define the structure and number of active hydrogen sites, or from titrimetric analysis, or determined empirically through experimentation to optimise cure/cured requirements.

A2.3.2 Blended and Filled Epoxy Formulations

To calculate the stoichiometric ratios where blends of resins of different epoxy equivalent weights are combined with blends of amines of different amine hydrogen equivalent weights, eqn (A2.3.5) is used to determine the EEW for a mix of epoxy functional resins/diluents (a/b) and eqn (A2.3.6) is used to determine the AHEW for a mix of amines (c/d) with active hydrogens.

$$\text{EEW of epoxy resin blend} = \frac{\text{Total Weight (Wt.) of resins in blend}}{\dfrac{\text{Wt.(a)}}{\text{EEW(a)}} + \dfrac{\text{Wt.(b)}}{\text{EEW(b)}}} \quad (A2.3.5)$$

$$\text{AHEW of an amine blend} = \frac{\text{Total Weight (Wt.) of amines in blend}}{\dfrac{\text{Wt.(c)}}{\text{AHEW(c)}} + \dfrac{\text{Wt.(d)}}{\text{AHEW(d)}}}$$

$$(A2.3.6)$$

Where there are no fillers or non-reactive components involved, the stoichiometric loading for the blends of epoxy resins and amine hardeners is then simply determined from the calculated blend EEW and blend AHEW values by the use of eqn (A2.3.3).

Typically, however, epoxy resin components of systems are formulated with non-reactive constituents, such as modifiers and solid fillers, so it is necessary to adjust the level of the curing agent to cure only the portion of the mix that is reactive *i.e.* the epoxy functional resins and reactive diluents. This is achieved by calculating the EEW of the formulated epoxy component from eqn (A2.3.7)

$$\text{EEW of formulated epoxy} = \frac{\text{Wt. epoxy blend} + \text{Wt. modifiers} + \text{Wt. fillers}}{\dfrac{\text{Wt. epoxy blend}}{\text{EEW epoxy blend}}}$$

$$(A2.3.7)$$

In the case where the amine component is formulated with non-reactive modifiers and fillers, the AHEW of the formulated amine is determined using eqn (A2.3.8)

$$\text{AHEW of formulated amine} = \frac{\text{Wt. amine blend} + \text{Wt. modifiers} + \text{Wt. fillers}}{\dfrac{\text{Wt. amine blend}}{\text{AHEW amine blend}}}$$

$$(A2.3.8)$$

and the loading of formulated amine, in Parts per Hundred by weight of formulated Resin, is calculated as before from eqn (A2.3.3).

A2.3.3 Worked Example

What quantities of graded aggregate, epoxy resin (EEW 190), epoxy diluent (EEW 145), amine hardener (AHEW 35), amine modifier (AHEW 115) and benzyl alcohol accelerator/plasticiser are required to make a 200 kg batch of epoxy mortar with an overall 8 : 1 *w/w* filler to binder ratio based on a 60/40 weight blend of epoxy resin/diluent, and 60/20/20 weight blend of amine hardener/modifier/benzyl alcohol, at nominal 100% stoichiometry?

Hint: Firstly establish the quantities of aggregate and binder, then determine the EEW for the epoxy blend and the AHEW for the amine blend. Next, calculate the loading requirement for the amine component, then work out how much epoxy blend and amine blend is required in the binder component before subdivision into the individual constituents.

i. 200 kg of an 8 : 1 *w/w* filler loaded mix would require $8/9 \times 200 = 178$ kg graded aggregate, and $1/9 \times 200 = 22$ kg binder.

ii. Using eqn (A2.3.5) the combined EEW for the epoxy component is

$$\text{EEW epoxy blend} = \frac{60 + 40}{\dfrac{60}{190} + \dfrac{40}{145}} = \frac{100}{0.315 + 0.276} = \frac{100}{0.591} = 169$$

iii. Using eqn (A2.3.6) the combined AHEW for the amine component is

$$\text{AHEW amine blend} = \frac{60 + 20 + 20}{\dfrac{60}{35} + \dfrac{20}{115}} = \frac{100}{1.714 + 01.74} = \frac{100}{1.888} = 53$$

iv. Therefore, the loading in PHR for the amine component using eqn (A2.3.3) is

$$\text{Amine blend PHR in epoxy blend} = \frac{53 \times 100}{169} = 31.4$$

v. So, 22 kg binder requires $100/(100 + 31.4) \times 22 = 16.74$ kg of epoxy blend comprising

$60/(60 + 40) \times 16.74 = 10.05$ kg epoxy resin (EEW 190), and

$40/(60 + 40) \times 16.74 = 6.69$ kg epoxy diluent (EEW 145)

and, $31.4/(100 + 31.4) \times 22 = 5.26$ kg of the amine blend which consists of

$60/(60 + 20 + 20) \times 5.26 = 3.16$ kg amine hardener

(AHEW 35), and

$20/(60 + 20 + 20) \times 5.26 = 1.05$ kg amine modifier

(AHEW 115), and

$20/(60 + 20 + 20) \times 5.26 = 1.05$ kg benzyl alcohol

accelerator/plasticiser

Answer:		
Graded Aggregate		178.00 kg
Epoxy resin (EEW 190)		10.05 kg
Epoxy diluent (EEW 145)		6.69 kg
Amine hardener (AHEW 35)		3.16 kg
Amine modifier (AHEW 115)		1.05 kg
Benzyl alcohol accelerator/plasticiser		1.05 kg
		200.00 kg

A2.4 Polyureas and Polyurethanes

Linear polyureas form when di-isocyanates are combined with long-chain amine-terminated polyether/polyester resins and short-chain diamine extenders, and linear polyurethanes form when di-isocyanates are combined with higher molecular weight (long-chain) diol and low-molecular-weight (short-chain) diol extenders. Three dimensionally crosslinked polyurethanes occur when one of the reaction constituents has three or more reactive sites—crosslinkers can be tri-isocyanates or multi-functional isocyanate prepolymers, or low-molecular-weight triols or high-molecular-weight polyols. In all cases, stoichiometric ratios of reactive components are critical to ensure controlled nucleophilic addition polymerisation, with specific catalysts necessary for their accelerating influence on the NCO/OH reaction. Any surplus isocyanate will gradually react with atmospheric humidity or undergo side reactions which increase crosslink density and hardness—less than stoichiometric quantities of isocyanate lead to under crosslinking and more flexibility. When formulating elastomers, coatings and plastics with polyurethanes, however, it is generally preferable to vary crosslinking/hardness/flexibility by maintaining stoichiometry with a poly-isocyanate and adjusting the selection of polyol—inclusion of additional higher hydroxyl content polyols for hardness and chemical resistance; incorporation of extra low hydroxyl content polyols for flexibility and softness.

A2.4.1 Basic Stoichiometry Calculations

Isocyanate Content or Isocyanate Value, expressed in weight percentage of reactive isocyanate groups (% NCO), for monomeric isocyanates of known molecular weight and functionality is calculated using eqn (A2.4.1), and for commercial mixtures of isocyanates, prepolymers and urethane intermediates it is measured by ASTM D2572 (no ISO equivalent) titrimetric analysis.

$$\% \, NCO = \frac{42 \times functionality \times 100}{Molecular \; Weight \; of \; isocyanate} \qquad (A2.4.1)$$

where $42 =$ the molecular weight of the NCO group.

Isocyanate Equivalent Weight (NCO EW) for monomeric isocyanates, prepolymers and urethane intermediates are then determined by reference to eqn (A2.4.2).

$$NCO \; equivalent \; weight = \frac{42 \times 100}{\% \, NCO} \qquad (A2.4.2)$$

Isocyanate Index (NCO Index) or Number, the amount of isocyanate to be used relative to the theoretical equivalent required critically important in stoichiometric ratio calculations, is calculated using eqn (A2.4.3).

$$NCO \; index = \frac{Actual \; amount \; of \; isocyanate \; used \times 100}{Theoretical \; amount \; of \; isocyanate \; required} \qquad (A2.4.3)$$

The desired stoichiometry for formulated polyureas and polyurethanes is achieved by determination of the total number of equivalents of the isocyanate required to react with each of the polyamine, polyol chain extender/crosslinker components, other reactive additives such as oxazolidones, and water in the case of foams. eqn (A2.4.4) is used to determine individual isocyanate equivalent requirement in Parts By Weight (PBW), inclusive of adjustments for NCO index, for each individual "A-side" component from which a total "B-side" requirement for isocyanate can be made by summation.

$$\text{PBW equivalents NCO} = \frac{\text{PBW reactive component} \times \text{NCO EW}}{\text{reactive component EW}} \times \frac{\text{NCO Index}}{100}$$

(A2.4.4)

The Equivalent Weights (EWs) of amines, alcohols and other reactive additives of known structure and fixed molecular weight can be calculated using eqn (A2.4.5).

$$\text{Equivalent Weight reactive component} = \frac{\text{Molecular Weight}}{\text{Functionality}}$$ (A2.4.5)

The EW for polyols with distributions of molecular weights is calculated using eqn (A2.4.6) following ISO 14900/ASTM D4274 titrimetric analysis of the OH number (hydroxyl number, the mg of KOH equivalent to the hydroxyl content in one gm of polyol or other hydroxyl compound) and the acid number (acid value for residual acidic material in the polyol reported in the same units as hydroxyl number).

$$\text{Polyol Equivalent Weight} = \frac{56.1 \times 1000}{\text{OH number} + \text{acid number}}$$ (A2.4.6)

where 56.1 is the molecular weight of KOH and 1000 is the number of mg in one gram of sample. Polyols with very low acid numbers are characterised by their OH number determined from eqn (A2.4.7).

$$\text{Polyol OH number} = \frac{56.1 \times 1000}{\text{polyol EW}}$$ (A2.4.7)

The EW for polyamines with distributions of molecular weights is calculated using eqn (A2.4.8) following ISO 25761/ASTM D6979 titrimetric analysis of the amine number (amine value, the mg of KOH equivalent in one gm of amine-functional polyols, polyether polyols and polyether polyol blends used in polyurethane reactions).

$$\text{Polyamine Equivalent Weight} = \frac{56.1 \times 1000}{\text{amine number}}$$ (A2.4.8)

Amine numbers for materials of fixed molecular weight can also be determined by calculation using eqn (A2.4.9).

$$\text{Amine Number} = \frac{56.1 \times 1000}{\text{amine Molecular Weight}} \qquad (A2.4.9)$$

A2.4.2 Blended and Filled Polyurethane Formulations

To calculate the stoichiometric ratios for blends of isocyanate pre-polymers and urethane intermediates (e and f, respectively) of different isocyanate equivalent weight then eqn (A2.4.10) is used to determine the net equivalent weight where there are no fillers or non-reactive components involved.

$$\text{NCO blend Equivalent Weight} = \frac{\text{Total Weight (Wt.) isocyanates in blend}}{\dfrac{\text{Wt. NCO(e)}}{\text{EW NCO(e)}} + \dfrac{\text{Wt. NCO(f)}}{\text{EW NCO(f)}}}$$

$$(A2.4.10)$$

Where the isocyanate component is formulated with non-reactive components such as plasticisers, moisture scavengers, pigments, thixotropes, catalysts, or inhibitors, eqn (A2.4.11) can be used to calculate the NCO EW for what is typically referred to as the "B-side".

Formulated B-side NCO EW

$$= \frac{\text{Wt. of NCO blend} + \text{Wt. modifiers} + \text{Wt. fillers}}{\dfrac{\text{Wt. of NCO blend}}{\text{EW of NCO blend}}} \qquad (A2.4.11)$$

In Section A2.4.1, it was indicated that eqn (A2.4.4) should be used to determine the number of equivalents of isocyanate required for each of the polyamine, polyol chain extender/crosslinker components, other reactive additives and water in the case of foams. This is a straightforward approach where the "A-side" is unformulated. However, where blends of amine with combinations of polyol extenders and crosslinkers (g and h, respectively) are formulated with non-reactive modifiers and fillers, the equivalent weight of a formulated "A-side" can be determined simply using eqn (A2.4.12).

Formulated A-side EW

$$= \frac{\text{Wt. reactive components} + \text{Wt. modifiers} + \text{Wt. fillers}}{\dfrac{\text{Wt. amine}}{\text{EW amine}} + \dfrac{\text{Wt. polyol(g)}}{\text{EW polyol(g)}} + \dfrac{\text{Wt. polyol(h)}}{\text{EW polyol(h)}}} \qquad (A2.4.12)$$

Alternatively, the equivalent weight for a polyol blend can be determined using eqn (A2.4.13).

$$\text{Polyol blend Equivalent Weight} = \frac{\text{Total Weight (Wt) of polyols in blend}}{\dfrac{\text{Wt. polyol(g)}}{\text{EW polyol(g)}} + \dfrac{\text{Wt. polyol(h)}}{\text{EW polyol(h)}}}$$

$$(A2.4.13)$$

Once the equivalent weights for the formulate "A-side" and "B-side" have been determined, eqn (A2.4.14) can be used to determine the formulated "B-side" isocyanate equivalent requirement in PBW for any given weight of formulated "A-side" blend inclusive of adjustments for NCO index.

$$\text{PBW equivalent NCO} = \frac{\text{PBW A-side} \times \text{B-side NCO EW}}{\text{A-side EW}} \times \frac{\text{NCO index}}{100}$$

$$(A2.4.14)$$

A2.4.3 Worked Example

What quantities of isocyanate prepolymer (3.5% NCO), isocyanate crosslinker (21% NCO), polyether diol (OH number 169), polyether triol (OH number 524), polyester polyol (OH number 160, acid number 0.2) and pigment/filler blend are required to make a 150 kg batch of concrete coating with an overall 0.8 : 1 *w/w* filler to binder ratio based on a 1/5 weight blend of isocyanate prepolymer/crosslinker and a 2/1/1 weight blend of polyether diol/triol/polyester polyol, at 105 NCO Index?

Hint: Firstly establish the quantities of pigment/filler blend and resin binder, then determine the NCO equivalent weight for the individual isocyanates and for the blend, then determine the equivalent weight of the polyol blend. Next, calculate the loading requirement for the polyol blend making an adjustment for targeted isocyanate index, and then work out how much isocyanate blend and polyol blend is required in the binder component before subdivision into to the individual constituents.

i. 150 kg of a 0.8 to 1 pigment/filler loaded mix would require $0.8/1.8 \times 150 = 67$ kg pigment/filler blend, and $1/1.8 \times 150 = 83$ kg binder.

ii. Using eqn (A2.4.2), the isocyanate equivalent weights for the isocyanate prepolymer (3.5% NCO) and isocyanate crosslinker (21% NCO) are:

$$\text{isocyanate prepolymer Equivalent Weight} = \frac{42 \times 100}{3.5} = 1200$$

$$\text{isocyanate crosslinker Equivalent Weight} = \frac{42 \times 100}{21} = 200$$

So, using eqn (A2.4.10), the 1/5 isocyanate blend equivalent weight is:

$$\text{NCO blend Equivalent Weight} = \frac{1+5}{1/1200 + 5/200} = 232$$

iii. Using eqn (A2.4.6), the polyol equivalent weights for the polyether diol (OH number 169), polyether triol (OH number 524), polyester polyol (OH number 160, acid number 0.2) are:

$$\text{Polyether diol Equivalent Weight} = \frac{56.1 \times 1000}{169} = 332$$

$$\text{Polyether triol Equivalent Weight} = \frac{56.1 \times 1000}{524} = 107$$

$$\text{Polyester polyol Equivalent Weight} = \frac{56.1 \times 1000}{160 + 0.2} = 350$$

So, using Equation (A2.4.13), the 2/1/1 polyol blend equivalent weight is:

$$\text{Polyol blend Equivalent Weight} = \frac{2+1+1}{\dfrac{2}{332} + \dfrac{1}{107} + \dfrac{1}{350}} = 220$$

iv. Using eqn (A2.4.4) to determine the isocyanate equivalent requirement in PBW for 100 PBW polyol blend, inclusive of adjustments for the 105 NCO index, gives:

$$\text{Isocyanate blend equivalent PBW requirement} = \frac{100 \times 232}{220} \times \frac{105}{100} = 111$$

v. So, 83 kg binder requires $111/(100+111) \times 83 = 43.67$ kg of isocyanate blend comprising

$$1/(1+5) \times 43.67 = 7.28 \text{ kg isocyanate prepolymer}$$

$$(3.5\% \text{ NCO}), \text{ and}$$

$$5/(1+5) \times 43.67 = 36.39 \text{ kg isocyanate crosslinker (21\% NCO)}$$

and, $100/(100+111) \times 83 = 39.33$ kg of the polyol blend which consists of

$$2/(2+1+1) \times 39.33 = 19.67 \text{ kg polyether diol}$$

$$(\text{OH number 169}), \text{ and}$$

$$1/(2+1+1) \times 39.33 = 9.83 \text{ kg polyether triol}$$

$$(\text{OH number 524}), \text{ and}$$

$$1/(2+1+1) \times 39.33 = 9.83 \text{ kg polyester polyol}$$

$$(\text{OH number 160, acid number 0.2})$$

Answer: Part A

Polyether diol (OH number 169)	19.67 kg
Polyether triol (OH number 524)	9.83 kg
Polyester polyol (OH number 160, acid number 0.2)	9.83 kg
Pigment/filler blend	67.00 kg

Part B

Isocyanate prepolymer (3.5% NCO)	7.28 kg
Isocyanate crosslinker (21% NCO)	36.39 kg
Total	150.00 kg

Appendix 3   Cited International Standards, Practises, Specifications and Test Methods

American National Standards Institute (ANSI), Washington DC, USA

ANSI/ESD S20.20	Standard for the Development of an Electrostatic Discharge Control Program for Protection of Electrical and Electronic Parts, Assemblies and Equipment (excluding Electrically Initiated Explosive Devices).
ANSI/UL 1709	Standard for Rapid Rise Fire Tests of Protection Materials for Structural Steel.

American Society for Testing and Materials (ASTM) International, Pennsylvania, USA

ASTM B117	Practice for Operating Salt Spray (Fog) Apparatus.
ASTM C33	Specification for Concrete Aggregates.
ASTM C39	Test Method for Compressive Strength of Cylindrical Concrete Specimens.
ASTM C78	Test Method for Flexural Strength of Concrete (Using Simple Beam with Third-Point Loading).

Industrial Polymer Applications: Essential Chemistry and Technology
By William R. Ashcroft
© William R. Ashcroft 2017
Published by the Royal Society of Chemistry, www.rsc.org

ASTM C109	Test Method for Compressive Strength of Hydraulic Cement Mortars (Using 2-in. or [50 mm] Cube Specimens).
ASTM C150	Specification for Portland cement.
ASTM C191	Test Methods for Time of Setting of Hydraulic Cement by Vicat Needle.
ASTM C267	Test Methods for Chemical Resistance of Mortars, Grouts, and Monolithic Surfacings and Polymer Concretes.
ASTM C297	Test Method for Flatwise Tensile Strength of Sandwich Constructions.
ASTM C309	Specification for Liquid Membrane-Forming Compounds for Curing Concrete.
ASTM C395	Specification for Chemical-Resistant Resin Mortars.
ASTM C580	Test Method for Flexural Strength and Modulus of Elasticity of Chemical-Resistant Mortars, Grouts, Monolithic Surfacings, and Polymer Concretes.
ASTM C719	Test Method for Adhesion and Cohesion of Elastomeric Joint Sealants under Cyclic Movement (Hockman Cycle).
ASTM C779	Test Method for Abrasion Resistance of Horizontal Concrete Surfaces.
ASTM C794	Test Method for Adhesion-in-Peel of Elastomeric Joint Sealants.
ASTM C868	Test Method for Chemical Resistance of Protective Linings.
ASTM C873	Test Method for Compressive Strength of Concrete Cylinders Cast in Place in Cylindrical Moulds.
ASTM C920	Specification for Elastomeric Joint Sealants.
ASTM C961	Test Method for Lap Shear Strength of Sealants.
ASTM C1202	Test Method for Electrical Indication of Concrete's Ability to Resist Chloride Ion Penetration.
ASTM C1305	Test Method for Crack Bridging Ability of Liquid-Applied Waterproofing Membrane.
ASTM C1306	Test Method for Hydrostatic Pressure Resistance of a Liquid-Applied Waterproofing Membrane.
ASTM C1315	Specification for Liquid Membrane-Forming Compounds Having Special Properties for Curing and Sealing Concrete.
ASTM C1438	Specification for Latex and Powder Polymer Modifiers for use in Hydraulic Cement Concrete and Mortar.

ASTM C1469	Test Method for Shear Strength of Joints of Advanced Ceramics at Ambient Temperature.
ASTM C1543	Test Method for Determining the Penetration of Chloride Ion into Concrete by Ponding.
ASTM C1549	Test Method for Determination of Solar Reflectance near Ambient Temperature Using a Portable Solar Reflectometer.
ASTM D149	Test Method for Dielectric Breakdown Voltage and Dielectric Strength of Solid Electrical Insulating Materials at Commercial Power Frequencies.
ASTM D150	Test Methods for AC Loss Characteristics and Permittivity (Dielectric Constant) of Solid Electrical Insulation.
ASTM D256	Test Methods for Determining the Izod Pendulum Impact Resistance of Plastics.
ASTM D257	Test Methods for DC Resistance or Conductance of Insulating Materials.
ASTM D395	Test Methods for Rubber Property—Compression Set.
ASTM D412	Test Methods for Vulcanized Rubber and Thermoplastic Elastomers—Tension.
ASTM D413	Test Methods for Rubber Property—Adhesion to Flexible Substrate.
ASTM D429	Test Methods for Rubber Property—Adhesion to Rigid Substrates.
ASTM D523	Test Method for Specular Gloss.
ASTM D543	Practices for Evaluating the Resistance of Plastics to Chemical Reagents.
ASTM D575	Test Methods for Rubber Properties in Compression.
ASTM D624	Test Method for Tear Strength of Conventional Vulcanized Rubber and Thermoplastic Elastomers.
ASTM D638	Test Method for Tensile Properties of Plastics.
ASTM D648	Test Method for Deflection Temperature of Plastics under Flexural Load in the Edgewise Position.
ASTM D660	Test Method for Evaluating Degree of Checking of Exterior Paints.
ASTM D695	Test Method for Compressive Properties of Rigid Plastics
ASTM D790	Test Methods for Flexural Properties of Unreinforced and Reinforced Plastics and Electrical Insulating Materials.

ASTM D897	Test Method for Tensile Properties of Adhesive Bonds.
ASTM D991	Test Method for Rubber Property—Volume Resistivity of Electrically Conductive and Antistatic Products.
ASTM D1002	Test Method for Apparent Shear Strength of Single-Lap-Joint Adhesively Bonded Metal Specimens by Tension Loading (Metal-to-Metal).
ASTM D1054	Test Method for Rubber Property-Resilience Using a Goodyear-Healey Rebound Pendulum (Withdrawn 2010, no replacement).
ASTM D1062	Test Method for Cleavage Strength of Metal-to-Metal Adhesive Bonds.
ASTM D1200	Test Method for Viscosity by Ford Viscosity Cup.
ASTM D1229	Test Method for Rubber Property-Compression Set at Low Temperatures.
ASTM D1308	Test Method for Effect of Household Chemicals on Clear and Pigmented Organic Finishes.
ASTM D1415	Test Method for Rubber Property—International Hardness.
ASTM D1525	Test Method for Vicat Softening Temperature of Plastics.
ASTM D1652	Test Method for Epoxy Content of Epoxy Resins.
ASTM D1654	Test Method for Evaluation of Painted or Coated Specimens Subjected to Corrosive Environments.
ASTM D1729	Practice for Visual Appraisal of Colours and Colour Differences of Diffusely-Illuminated Opaque Materials.
ASTM D1876	Test Method for Peel Resistance of Adhesives (T-Peel Test).
ASTM D2095	Test Method for Tensile Strength of Adhesives by Means of Bar and Rod Specimens.
ASTM D2197	Test Method for Adhesion of Organic Coatings by Scrape Adhesion.
ASTM D2228	Test Method for Rubber Property-Relative Abrasion Resistance by the Pico Abrader Method.
ASTM D2240	Test Method for Rubber Property—Durometer Hardness.
ASTM D2485	Test Methods for Evaluating Coatings for High Temperature Service.
ASTM D2572	Test Method for Isocyanate Groups in Urethane Materials or Prepolymers.

ASTM D2632	Test Method for Rubber Property—Resilience by Vertical Rebound.
ASTM D2794	Test Method for Resistance of Organic Coatings to the Effects of Rapid Deformation (Impact).
ASTM D2803	Guide for Testing Filiform Corrosion Resistance of Organic Coatings on Metal.
ASTM D3039	Test Method for Tensile Properties of Polymer Matrix Composite Materials.
ASTM D3164	Test Method for Strength Properties of Adhesively Bonded Plastic Lap-Shear Sandwich Joints in Shear by Tension Loading.
ASTM D3166	Test Method for Fatigue Properties of Adhesives in Shear by Tension Loading (Metal/Metal).
ASTM D3322	Practice for Testing Primers and Primer Surfacers over Preformed Metal.
ASTM D3359	Test Methods for Measuring Adhesion by Tape Test.
ASTM D3574	Test Methods for Flexible Cellular Materials—Slab, Bonded, and Moulded Urethane Foams.
ASTM D3623	Test Method for Testing Antifouling Panels in Shallow Submergence.
ASTM D3762	Test Method for Adhesive-Bonded Surface Durability of Aluminium (Wedge Test).
ASTM D3807	Test Method for Strength Properties of Adhesives in Cleavage Peel by Tension Loading (Engineering Plastics-to-Engineering Plastics).
ASTM D4060	Test Method for Abrasion Resistance of Organic Coatings by the Taber Abraser.
ASTM D4214	Test Methods for Evaluating the Degree of Chalking of Exterior Paint Films.
ASTM D4263	Test Method for Indicating Moisture in Concrete by the Plastic Sheet Method.
ASTM D4274	Test Methods for Testing Polyurethane Raw Materials: Determination of Hydroxyl Numbers of Polyols.
ASTM D4501	Test Method for Shear Strength of Adhesive Bonds between Rigid Substrates by the Block-Shear Method.
ASTM D4541	Test Method for Pull-Off Strength of Coatings Using Portable Adhesion Testers.
ASTM D4585	Practice for Testing Water Resistance of Coatings Using Controlled Condensation.

ASTM D4762	Guide for Testing Polymer Matrix Composite Materials.
ASTM D4938	Test Method for Erosion Testing of Antifouling Paints Using High Velocity Water.
ASTM D4939	Test Method for Subjecting Marine Antifouling Coating to Biofouling and Fluid Shear Forces in Natural Seawater.
ASTM D5045	Test Methods for Plane-Strain Fracture Toughness and Strain Energy Release Rate of Plastic Materials.
ASTM D5083	Test Method for Tensile Properties of Reinforced Thermosetting Plastics Using Straight-Sided Specimens.
ASTM D5108	Test Method for Organotin Release Rates of Antifouling Coating Systems in Sea Water.
ASTM D5363	Standard for Anaerobic Single-Component Adhesives (AN)
ASTM D5385	Test Method for Hydrostatic Pressure Resistance of Waterproofing Membranes.
ASTM D5379	Test Method for Shear Properties of Composite Materials by the V-Notched Beam Method.
ASTM D5401	Test Method for Evaluating Clear Water Repellent Coatings on Wood.
ASTM D5479	Practice for Testing Biofouling Resistance of Marine Coatings Partially Immersed.
ASTM D5499	Test Methods for Heat Resistance of Polymer Linings for Flue Gas Desulfurization Systems.
ASTM D5618	Test Method for Measurement of Barnacle Adhesion Strength in Shear.
ASTM D6110	Test Method for Determining the Charpy Impact Resistance of Notched Specimens of Plastics.
ASTM D6442	Test Method for Determination of Copper Release Rate from Antifouling Coatings in Substitute Ocean Water.
ASTM D6671	Test Method for Mixed Mode I-Mode II Interlaminar Fracture Toughness of Unidirectional Fibre Reinforced Polymer Matrix Composites.
ASTM D6677	Test Method for Evaluating Adhesion by Knife.
ASTM D6695	Practice for Xenon-Arc Exposures of Paint and Related Coatings.
ASTM D6979	Test Method for Polyurethane Raw Materials: Determination of Basicity in Polyols, Expressed as Percent Nitrogen.

ASTM D6903	Test Method for Determination of Organic Biocide Release Rate from Antifouling Coatings in Substitute Ocean Water.
ASTM D6904	Practice for Resistance to Wind-Driven Rain for Exterior Coatings Applied on Masonry.
ASTM D6943	Practice for Immersion Testing of Industrial Protective Coatings and Linings.
ASTM D7234	Test Method for Pull-Off Adhesion Strength of Coatings on Concrete Using Portable Pull-Off Adhesion Testers.
ASTM D7264	Test Method for Flexural Properties of Polymer Matrix Composite Materials.
ASTM D7832	Standard Guide for Performance Attributes of Waterproofing Membranes Applied to Below-Grade Walls/Vertical Surfaces (Enclosing Interior Spaces).
ASTM E84	Test Method for Surface Burning Characteristics of Building Materials.
ASTM E96	Test Methods for Water Vapour Transmission of Materials.
ASTM E111	Test Method for Young's Modulus, Tangent Modulus, and Chord Modulus.
ASTM E119	Test Methods for Fire Tests of Building Construction and Materials.
ASTM E303	Test Method for Measuring Surface Frictional Properties Using British Pendulum Tester.
ASTM E514	Test Method for Water Penetration and Leakage through Masonry.
ASTM E831	Test Method for Linear Thermal Expansion of Solid Materials by Thermomechanical Analysis.
ASTM E903	Test Method for Solar Absorptance, Reflectance, and Transmittance of Materials Using Integrating Spheres.
ASTM E1004	Test Method for Determining Electrical Conductivity Using the Electromagnetic (Eddy-Current) Method.
ASTM E1131	Test Method for Compositional Analysis by Thermogravimetry.
ASTM E1164	Practice for Obtaining Spectrometric Data for Object-Colour Evaluation.
ASTM E1980	Practice for Calculating Solar Reflectance Index of Horizontal and Low-Sloped Opaque Surfaces.
ASTM E2058	Test Methods for Measurement of Material Flammability Using a Fire Propagation Apparatus.

ASTM F433	Practice for Evaluating Thermal Conductivity of Gasket Materials.
ASTM G6	Test Method for Abrasion Resistance of Pipeline Coatings.
ASTM G8	Test Methods for Cathodic Disbonding of Pipeline Coatings.
ASTM G20	Test Method for Chemical Resistance of Pipeline Coatings.
ASTM G32	Test Method for Cavitation Erosion Using Vibratory Apparatus.
ASTM G42	Test Method for Cathodic Disbonding of Pipeline Coatings Subjected to Elevated Temperatures.
ASTM G65	Test Method for Measuring Abrasion Using the Dry Sand/Rubber Wheel Apparatus.
ASTM G73	Test Method for Liquid Impingement Erosion Using Rotating Apparatus.
ASTM G76	Test Method for Conducting Erosion Tests by Solid Particle Impingement Using Gas Jets.
ASTM G85	Practice for Modified Salt Spray (Fog) Testing A5, dilute electrolyte cyclic fog dry test.
ASTM G151	Practice for Exposing Non-metallic Materials in Accelerated Test Devices that Use Laboratory Light Sources
ASTM G154	Practice for Operating Fluorescent Ultraviolet (UV) Lamp Apparatus for Exposure of Non-metallic Materials.
ASTM G155	Practice for Operating Xenon Arc Light Apparatus for Exposure of Non-Metallic Materials
ASTM G189	Guide for Laboratory Simulation of Corrosion-Under-Insulation.

British Standards Institute (BSI), London, UK

BS 476-3	Fire tests on building materials and structures. Classification and method of test for external fire exposure to roofs.
BS 476-4	Fire tests on building materials and structures. Non-combustibility test for materials.
BS 476-6	Fire tests on building materials and structures. Method of test for fire propagation for products.

BS 476-7	Fire tests on building materials and structures. Method of test to determine the classification of the surface spread of flame of products.
BS 476-20	Fire tests on building materials and structures. Method for determination of the fire resistance of elements of construction (general principles).
BS 476-21	Fire tests on building materials and structures. Methods for determination of the fire resistance of loadbearing elements of construction.
BS 476-22	Fire tests on building materials and structures. Method for determination of the fire resistance of non-loadbearing elements of construction.
BS 3900-D10	Methods of test for paints. Determination of colour and colour difference.
BS 4550	Methods of testing cement. Physical tests. Test for setting times.
BS 5350-C1	Methods of test for adhesives. Adhesively bonded joints: mechanical tests. Determination of cleavage strength of adhesive bonds.
BS 5350-C5	Methods of test for adhesives. Adhesively bonded joints: mechanical tests. Determination of bond strength in longitudinal shear.
BS 7079	General introduction to standards for preparation of steel substrates before application of paints and related products.
BS 7976-2	Pendulum testers. Method of operation.
BS 8102	Code of practice for protection of below ground structures against water from the ground.
BS 8204	Screeds, bases and *in situ* floorings. Concrete bases and cementitious levelling screeds to receive floorings. Code of practice.
BS EN 197	Cement. Composition, specifications and conformity criteria for common cements
BS EN 1062	Paints and varnishes. Coating materials and coating systems for exterior masonry and concrete. Classification.
BS EN 1062-7	Paints and varnishes. Coating materials and coating systems for exterior masonry and concrete. Determination of crack bridging properties.

BS EN 1297	Flexible sheets for waterproofing. Bitumen, plastic and rubber sheets for roof waterproofing. Method of artificial ageing by long term exposure to the combination of UV radiation, elevated temperature and water.
BS EN 1465	Adhesives. Determination of tensile lap-shear strength of bonded assemblies.
BS EN 1607	Thermal insulating products for building applications. Determination of tensile strength perpendicular to faces.
BS EN 1928	Flexible sheets for waterproofing. Bitumen, plastic and rubber sheets for roof waterproofing. Determination of watertightness.
BS EN 1931	Flexible sheets for waterproofing. Bitumen, plastic and rubber sheets for roof waterproofing. Determination of water vapour transmission properties.
BS EN 12190	Products and systems for the protection and repair of concrete structures. Test methods. Determination of compressive strength of repair mortar.
BS EN 12390	Testing hardened concrete. Shape, dimensions and other requirements for specimens and moulds.
BS EN 13381-8	Test methods for determining the contribution to the fire resistance of structural members. Applied reactive protection to steel members.
BS EN 12617	Products and systems for the protection and repair of concrete structures. Test methods. Determination of linear shrinkage for polymers and surface protection systems.
BS EN 12618-1	Products and systems for the protection and repair of concrete structures. Test methods. Adhesion and elongation capacity of injection products with limited ductility
BS EN 12637-1	Products and systems for the protection and repair of concrete structures. Test methods. Compatibility of injection products. Compatibility with concrete.
BS EN 12691	Flexible sheets for waterproofing. Bitumen, plastic and rubber sheets for roof waterproofing. Determination of resistance to impact.
BS EN 12730	Flexible sheets for waterproofing. Bitumen, plastic and rubber sheets for roof waterproofing. Determination of resistance to static loading.

BS EN 13036-4	Road and airfield surface characteristics. Test methods. Method for measurement of slip/skid resistance of a surface: The pendulum test.
BS EN 13813	Screed material and floor screeds. Screed material. Properties and requirements.
BS EN 13892-2	Methods of test for screed materials. Determination of flexural and compressive strength.
BS EN 13948	Flexible sheets for waterproofing. Bitumen, plastic and rubber sheets for roof waterproofing. Determination of resistance to root penetration.
BS EN 15771	Vitreous and porcelain enamels. Determination of surface scratch hardness according to the Mohs scale.
BS EN 15870	Adhesives. Determination of tensile strength of butt joints.

European Organisation for Technical Assessment (EOTA), Brussels, Belgium

EOTA TR005	Determination of the resistance to wind loads of partially bonded roof waterproofing membranes.
EOTA TR006	Determination of the resistance to dynamic indentation.
EOTA TR007	Determination of the resistance to static indentation.
EOTA TR008	Determination of the resistance to fatigue movement.
EOTA TR010	Exposure procedure for artificial weathering.
EOTA TR011	Exposure procedure for accelerated ageing by heat.
EOTA TR012	Exposure procedure for accelerated ageing by hot water.

International Electrotechnical Commission (IEC), Geneva, Switzerland

IEC 60085	Electrical insulation – Thermal evaluation and designation.
IEC 60093	Methods of test for volume resistivity and surface resistivity of solid electrical insulating materials.

IEC 60243	Electric strength of insulating materials – Test methods.
IEC 60250	Recommended methods for the determination of the permittivity and dielectric dissipation factor of electrical insulating materials at power, audio and radio frequencies including metre wavelengths.
IEC 61340-1	Electrostatics – Part 1: Electrostatic phenomena – Principles and measurements.

International Organization for Standardization (ISO), Geneva, Switzerland

ISO 34-2	Rubber, vulcanized or thermoplastic—Determination of tear strength—Part 2: Small (Delft) test pieces.
ISO 37	Rubber, vulcanized or thermoplastic. Determination of tensile stress-strain properties.
ISO 48	Rubber, vulcanized or thermoplastic– Determination of hardness (hardness between 10 IRHD and 100 IRHD).
ISO 75-1,2	Plastics – Determination of temperature of deflection.
ISO 75-3	Plastics – Determination of temperature of deflection under load – Part 3: High-strength thermosetting laminates and long-fibre-reinforced plastics.
ISO 175	Plastics – Methods of test for the determination of the effects of immersion in liquid chemicals.
ISO 178	Plastics – Determination of flexural properties.
ISO 179	Plastics – Determination of Charpy impact properties.
ISO 180	Plastics – Determination of Izod impact strength.
ISO 527	Plastics – Determination of tensile properties.
ISO 604	Plastics – Determination of compressive properties.
ISO 815-1,2	Rubber, vulcanized or thermoplastic – Determination of compression set.
ISO 868	Plastics and ebonite – Determination of indentation hardness by means of a durometer (Shore hardness)
ISO 1629	Rubber and latices – Nomenclature.
ISO 1827	Rubber, vulcanized or thermoplastic– Determination of shear modulus and adhesion to rigid plates – Quadruple-shear methods.

ISO 2431	Paints and varnishes – Determination of flow time by use of flow cups.
ISO 2813	Paints and varnishes – Determination of gloss value at 20 degrees, 60 degrees and 85 degrees.
ISO 4624	Paints and varnishes – Pull-off test for adhesion.
ISO 4628-1	Paints and varnishes – Evaluation of degradation of coatings – Designation of quantity and size of defects, and of intensity of uniform changes in appearance – Part 1: General introduction and designation system
ISO 4628-2	Paints and varnishes – Evaluation of degradation of coatings – Designation of quantity and size of defects, and of intensity of uniform changes in appearance – Part 2: Assessment of degree of blistering.
ISO 4628-3	Paints and varnishes – Evaluation of degradation of coatings – Designation of quantity and size of defects, and of intensity of uniform changes in appearance – Part 3: Assessment of degree of rusting.
ISO 4628-4	Paints and varnishes – Evaluation of degradation of coatings – Designation of quantity and size of defects, and of intensity of uniform changes in appearance – Part 4: Assessment of degree of cracking.
ISO 4628-5	Paints and varnishes – Evaluation of degradation of coatings – Designation of quantity and size of defects, and of intensity of uniform changes in appearance – Part 5: Assessment of degree of flaking.
ISO 4628-6	Paints and varnishes – Evaluation of degradation of coatings – Designation of quantity and size of defects, and of intensity of uniform changes in appearance – Part 6: Assessment of degree of chalking by tape method.
ISO 4628-7	Paints and varnishes – Evaluation of degradation of coatings – Designation of quantity and size of defects, and of intensity of uniform changes in appearance – Part 7: Assessment of degree of chalking by velvet method.
ISO 4628-8	Paints and varnishes – Evaluation of degradation of coatings – Designation of quantity and size of defects, and of intensity of uniform changes in appearance – Part 8: Assessment of degree of delamination and corrosion around a scribe or other artificial defect.

ISO 4628-10	Paints and varnishes – Evaluation of degradation of coatings – Designation of quantity and size of defects, and of intensity of uniform changes in appearance – Part 10: Assessment of degree of filiform corrosion
ISO 4649	Rubber, vulcanized or thermoplastic– Determination of abrasion resistance using a rotating cylindrical drum device.
ISO 4662	Rubber, vulcanized or thermoplastic– Determination of rebound resilience.
ISO 5470-1	Rubber- or plastics-coated fabrics – Determination of abrasion resistance – Part 1: Taber abrader.
ISO 5470-2	Rubber- or plastics-coated fabrics – Determination of abrasion resistance – Part 2: Martindale abrader.
ISO 6270-1	Paints and varnishes – Determination of resistance to humidity – Part 1: Continuous condensation.
ISO 6270-2	Paints and varnishes – Determination of resistance to humidity – Part 2: Procedure for exposing test specimens in condensation-water atmospheres.
ISO 6272-1,2	Paints and varnishes – Rapid-deformation (impact resistance) tests.
ISO 6707-1	Buildings and civil engineering works – Vocabulary – Part 1: General terms.
ISO 6721-11	Plastics – Determination of dynamic mechanical properties – Part 11: Glass transition temperature.
ISO 7031	Concrete hardened - Determination of the depth of penetration of water under pressure.
ISO 7619-1,2	Rubber, vulcanized or thermoplastic – Determination of indentation hardness methods.
ISO 7267-1,2	Rubber-covered rollers – Determination of apparent hardness methods.
ISO 7783	Paints and varnishes – Determination of water-vapour transmission properties – Cup method.
ISO 8501	Preparation of steel substrates before application of paints and related products – Visual assessment of surface cleanliness.
ISO 8502-2	Preparation of steel substrates before application of paints and related products – Tests for the assessment of surface cleanliness – Part 2: Laboratory determination of chloride on cleaned surfaces.

ISO 8510-2	Adhesives – Peel test for a flexible-bonded-to-rigid test specimen assembly – Part 2: 180 degree peel.
ISO 11339	Adhesives – T-peel test for flexible-to-flexible bonded assemblies.
ISO 11357-2	Plastics – Differential scanning calorimetry (DSC) – Part 2: Determination of glass transition temperature and glass transition step height.
ISO 11358-1-3	Plastics – Thermogravimetry (TG) of polymers.
ISO 11359-1-3	Plastics – Thermomechanical analysis (TMA).
ISO 11600	Building construction – Jointing products – Classification and requirements for sealants.
ISO 11997-1	Paints and varnishes – Determination of resistance to cyclic corrosion conditions – Part 1: Wet (salt fog)/dry/humidity.
ISO 11997-2	Paints and varnishes – Determination of resistance to cyclic corrosion conditions – Part 2: Wet (salt fog)/dry/humidity/UV light.
ISO 12944	Paints and varnishes – Corrosion protection of steel structures by protective paint systems.
ISO 13586	Plastics – Determination of fracture toughness (GIC and KIC) – Linear elastic fracture mechanics (LEFM) approach.
ISO 14900	Plastics – Polyols for use in the production of polyurethane – Determination of hydroxyl number.
ISO 15106	Plastics – Film and sheeting – Determination of water vapour transmission rate methods.
ISO 15235	Preparation of steel substrates before application of paints and related products – Collected information on the effect of levels of water-soluble salt contamination.
ISO 15695	Vitreous and porcelain enamels – Determination of scratch resistance of enamel finishes.
ISO 15711	Paints and varnishes – Determination of resistance to cathodic disbonding of coatings exposed to sea water.
ISO 16773	Paints and varnishes – Electrochemical impedance spectroscopy (EIS) on high-impedance coated specimens.
ISO 17463	Paints and varnishes – Guidelines for the determination of anticorrosive properties of organic coatings by accelerated cyclic electrochemical technique.

ISO 20340	Paints and varnishes – Performance requirements for protective paint systems for offshore and related structures.
ISO 22899	Determination of the resistance to jet fires of passive fire protection materials.
ISO 25761	Plastics – Polyols for use in the production of polyurethanes – Determination of basicity (total amine value), expressed as percent nitrogen.
ISO 28706	Vitreous and porcelain enamels – Determination of resistance to chemical corrosion.
ISO 28721	Vitreous and porcelain enamels – Glass-lined apparatus for process plants.

NACE International, Texas, USA (formerly National Association of Corrosion Engineers)

NACE No. 1	White Metal Blast Cleaning.
NACE No. 2	Near-White Metal Blast Cleaning.
NACE No. 3	Commercial Blast Cleaning.
NACE No. 4	Brush-Off Blast Cleaning.
NACE No. 6	Surface Preparation of Concrete.
NACE No. 8	Industrial Blast Cleaning.
NACE WJ-1	Waterjet Cleaning Of Metals – Clean To Bare Substrate.
NACE WJ-2	Waterjet Cleaning of Metals – Very Thorough Cleaning.
NACE WJ-3	Waterjet Cleaning of Metals – Thorough Cleaning.
NACE WJ-4	Waterjet Cleaning of Metals – Light Cleaning.
SP0198	Control of Corrosion under Thermal Insulation and Fireproofing Materials.
TM0174	Laboratory Methods for the Evaluation of Protective Coatings and Lining Materials on Metallic Substrates in Immersion Service.
TM0185	Evaluation of Internal Plastic Coatings for Corrosion Control of Tubular Goods by Autoclave Testing.

Norsok (Norsk Sokkels Konkuranseposisjon), Norway

| Norsok M-501 | Surface preparation and protective coating. |

Steel Structures Painting Council (now Society for Protective Coatings), Pittsburgh, USA

SSPC-SP1	Solvent cleaning.
SSPC-SP2	Hand tool cleaning.
SSPC-SP3	Power tool cleaning.
SSPC-SP5	White metal blast cleaning.
SSPC-SP6	Commercial Blast Cleaning.
SSPC-SP7	Brush-off blast cleaning.
SSPC-SP10	Near-white metal blast cleaning.
SSPC-SP13	Surface Preparation of Concrete.
SSPC-SP14	Industrial blast cleaning.
SSPC-SP WJ-1	Waterjet Cleaning Of Metals – Clean To Bare Substrate.
SSPC-SP WJ-2	Waterjet Cleaning of Metals – Very Thorough Cleaning.
SSPC-SP WJ-3	Waterjet Cleaning of Metals – Thorough Cleaning.
SSPC-SP WJ-4	Waterjet Cleaning of Metals – Light Cleaning.

Subject Index